Structural Integrity and Materials in Nuclear Power Plants

Structural Integrity and Materials in Nuclear Power Plants

the Proceedings of

TAGSI–FESI Symposium 2018

held on

18th April 2018

CAMBRIDGE, UK

COPYRIGHT © 2019, BY FESI PUBLISHING (FORMERLY EMAS PUBLISHING) AND © 2019, BY TAGSI
ALL RIGHTS RESERVED.
NO PART OF THIS DOCUMENT MAY BE REPRODUCED, STORED IN A RETRIEVAL SYSTEM, OR TRANSMITTED IN ANY FORM OR BY ANY MEANS, ELECTRONIC, MECHANICAL, PHOTOCOPYING, RECORDING OR OTHERWISE, WITHOUT THE PRIOR WRITTEN PERMISSION OF THE COPYRIGHT HOLDER, EMAS PUBLISHING.
THE CHAPTERS ARE REPRODUCED BY PERMISSION OF THE INDIVIDUAL AUTHORS.
APPLICATIONS FOR THE COPYRIGHT HOLDER'S WRITTEN PERMISSION SHOULD BE ADDRESSED THROUGH THE PUBLISHERS, FESI PUBLISHING.

FESI PUBLISHING
Suite 9, Derby House Chambers, Lytham Road, Fulwood, Preston PR2 8JE, UK
Tel. +44(0)1772 716786
fesipublishing@fesi.org.uk
www.fesipublishing.org.uk

ISBN: 978-0-9935485-1-2

PRINTED AND BOUND IN THE UNITED KINGDOM BY DXG MEDIA LTD
www.dxgmedia.com

DESIGN BY ANDREW BELL
andrewbell3005@gmail.com

TYPESET IN MODERN EXTENDED NO. 1, a characteristic English typeface released by the Monotype Studio in 1902. The cover shows a computer-generated aerial view of Hinkley Point C nuclear power station, Somerset, UK (Hydrock, CC BY-SA 4.0).

Preface

This symposium was in the series of TAGSI – FESI joint symposia held since 2010. This particular symposium, entitled "Structural Integrity and Materials in Nuclear Power Plants", was dedicated to the memory of Professor John Knott, OBE, FRS, FREng. John died in 2017 and was both a chair of TAGSI and a director of FESI. He contributed very positively over a number of years to the work of both groups. Indeed it was John's vision that prompted and promoted the need for, and the importance of, engineering structural integrity as a multi-disciplinary subject.

John was a leading expert in the fields of fracture and structural integrity in relation particularly to nuclear power generation plants. But more generally he made significant contributions to the quantitative scientific understanding of fracture processes in metals and alloys and applications to engineering. His early work was devoted to addressing the role of microstructure in the initiation and propagation of cracks, including the micro-mechanisms of cleavage, ductile and fatigue fracture in steels and non-ferrous alloys. Indeed he was the author of a much used text book on the principles of fracture mechanics (available as a free download on the FESI website: www.fesi.org.uk).

As a consequence of John's work related to nuclear power generation the papers in this symposium, written by experts in the field, illustrate the wide range of understanding and developments that have occurred within recent years to secure safer operation of nuclear power plants. Indeed several of the authors received training and guidance from John. We feel sure that the comprehensive coverage offered by these papers would have received full support from John. But no doubt, had he been present, he would have been challenged the authors in his incisive but kindly way.

R. A. AINSWORTH AND P. E. J. FLEWITT.

May, 2019.

John Frederick Knott (1938–2017)

Contents

	PAGE
PREFACE,	v

Mechanistic Aspects of the Damage-Tolerance of Advanced Multiple-Element Metallic Alloys and the Legacy of John Knott. — By Prof. Robert O. Ritchie, *University of California, Berkeley*, ... 1

Assessment of Non-Sharp Defects Using the Notch Failure Assessment Diagram. — By Anthony J. Horn, *Wood plc*, ... 9

Creep–Fatigue Crack Initiation in Weldments. — By Marc J. Chevalier and David W. Dean, *EDF Energy*, ... 23

Strengths and Weaknesses of the Proof Pressure Test Argument in RPV Structural Integrity Assessments: a TAGSI View. — By Prof. Stephen J. Garwood, *Imperial College London*, ... 36

Stress Based NDE: Taking Infrared Thermography Inspection from the Laboratory to the Power Station. — By Rachael C. Tighe, *University of Waikato*, J. Philip Tyler, *Enabling Process Technologies*, Geoffrey P. Howell, and Prof. Janice M. Dulieu-Barton, *University of Southampton*, ... 59

Probabilistic Structural Integrity. — By Nader A. Zentuti, Prof. Julian D. Booker, Joshua Hoole, Richard A. W. Bradford, and Prof. David Knowles, *University of Bristol*, ... 68

Generic Design Assessment and Structural Integrity Challenges: Past, Present, and Future? — By Jim P. Caul, *Office for Nuclear Regulation*, ... 80

Mechanistic Aspects of the Damage-Tolerance of Advanced Multiple-Element Metallic Alloys and the Legacy of John Knott

R. O. RITCHIE

ABSTRACT.—There are but a few individuals who have left a greater legacy in education and research in the field of fracture than John Knott, who was known throughout the world for his wit and wisdom in this critical aspect of science and engineering. As a tribute to his impact in the field of mechanical metallurgy, I would like to describe in this presentation some of the new structural materials that have emerged over the past decade or so, which involve multiple-element compositions. Here I refer primarily to so-called *high-entropy alloys*, which in principle are single-phase solid solutions comprising equiatomic compositions of five elements or more. Some of these new alloys, specifically those based on CrCoNi compositions, display some of the highest damage-tolerance, in terms of unprecedented combinations of strength and toughness, reported to date. The nano-scale origins of these properties are examined using in situ fracture testing in the transmission electron microscope and involve a duality of dislocation mechanisms, sometimes coupled with deformation twinning, which act to promote both strength and ductility, particularly at cryogenic temperatures.

I. Introduction

Professor John Frederick Knott (Fig. 1) has left an immense 'footprint' in the field of structural metallurgy, spanning the full spectrum of fracture and fatigue research, from the practical implementation of continuum fracture mechanics to the mechanistic identification on nano- and micro-scale fracture mechanisms. Although he is perhaps best known in the TAGSI/FESI communities for his tireless work in assessing and maintaining the structural integrity of real structures—pressure vessels, nuclear reactors, and the like—to my mind, his might in our field resulted primarily from his ability to couple such continuum-scale structural engineering to a lucid nano- to micro-scale understanding of the salient mechanisms involved. Starting with his days in the 1950s and 1960s as a graduate student in Cambridge under Alan Cottrell to his time at the Leatherhead CEGB Labs under Ted Smith where he was at the forefront of the mechanics and mechanisms of notched-bar cleavage fracture, to his initial time back in Cambridge (when I was his student) where his focus for the first few years was on ductile fracture and the development of CTOD testing, John maintained this balance, in essence between engineering and metallurgy. At that time in the late 1950s to the early 1970s, linear-elastic fracture mechanics was being developed, largely in the United States, but essentially as solely

---

Materials Sciences Division, Lawrence Berkeley National Laboratory, and Department of Materials Science & Engineering, University of California, Berkeley, CA 94720, US.

Fig. 1. — Professor John Knott in his office at Birmingham University, where he was Dean of Engineering. The date of this photograph is uncertain but the nature of his state of the art computer probably gives it away!

an applied mechanics endeavour. When Jim Rice spent his sabbatical in Cambridge in 1971–72, this union of thought truly came to its fruition; Rice and mechanics colleagues were pursuing the relatively new field of 'micromechanics' by venturing below the continuum, which perfectly coupled with John's understanding of the micro-scale mechanistic processes involved. Indeed, when I successfully nominated him as a Foreign Member of the US National Academy of Engineering in 2002, I (perhaps facilely) referred to him as the 'Father of Microstructural Fracture Mechanics'.

I have tried to shape my own academic approach to the field of fracture exactly in the manner that John did, and so along with his many other students, I hope in some small way that I can represent part of his vast legacy. This is pertinent as in recent years the field of structural metallurgy has enjoyed somewhat of a revival through the introduction, into the research world at least, of several new classes of structural metallic alloys — here I refer to numerous multiple-element alloys in the form of bulk metallic glasses and now high-entropy alloys. Although research on these new materials has readily populated the storied pages of *Nature* and *Science*, none of these alloys have to date been used much in real life, full-scale engineering applications.* But why? Well one obvious reason is that, unlike electronic materials and devices, the time between the development of structural materials in the laboratory and their translation to actual engineering application is not measured in months but invariably in decades (witness how long it took for polymer composites to be used in commercial airframes and now ceramic composites in commercial gas turbine engines). The other reason is that to develop such alloys for actual use takes the synergistic mechanistic and structural engineering approach that John Knott always advocated, in order to create such a translation from concept to reality. Furthermore, to delve into the details, it is seldom trivial to create the vital property of damage-tolerance in new structural materials.

* To be fair, bulk metallic glass alloys have been used to make tiny SIM card holders in recent variants of the iPhone!

Damage-tolerance can be an elusive characteristic in structural alloys as it requires both strength and ductility, properties that are more often than not mutually exclusive. However, certain new equiatomic, multiple-element metallic alloys—termed *high-entropy alloys*—are of interest here. The name refers to the hypothesis that if an equiatomic alloy comprises five or more elements, the configuration entropy should overwhelm the enthalpy of phase formation to result in a solid solution. Although clearly this is not always the case, the so-called Cantor CrMnFeCoNi alloy is a true high-entropy alloy. CrMnFeCoNi and its derivatives, such as the medium-entropy CrCoNi alloy, are single-phase, face-centred cubic (fcc) solid solutions which can display exceptional mechanical properties, with strengths in excess of 1 GPa and ductilities of ~60–90%, which result in outstanding fracture toughness values ($K_{JIc} > 200$ MPa m); moreover, these properties are further enhanced at cryogenic temperatures (*1*, *2*).

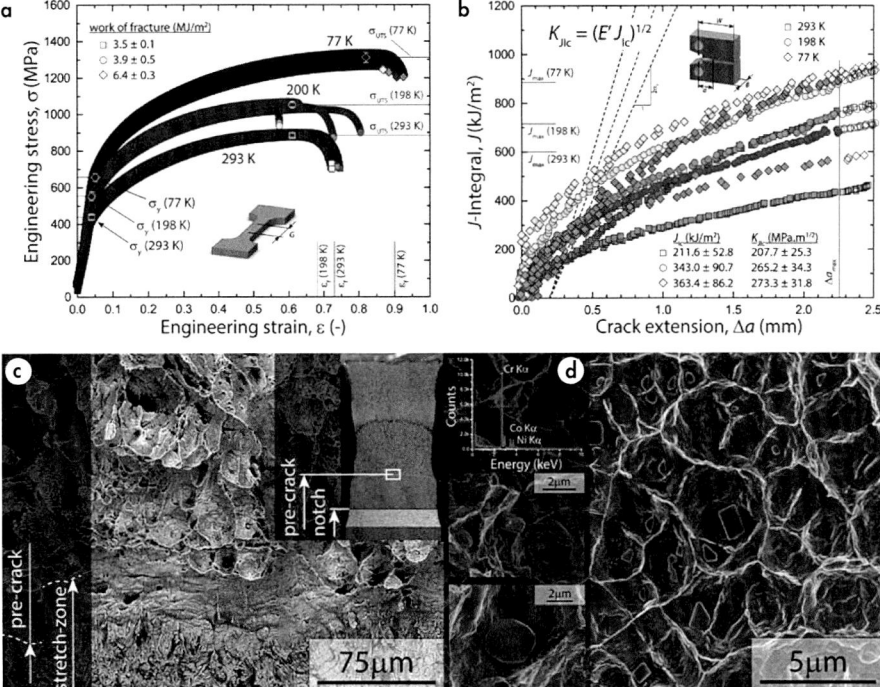

FIG. 2. — Mechanical properties and failure characteristics of CrCoNi medium-entropy alloy. *a*. Tensile tests show a significant increase in yield strength, σ_y, tensile strength, σ_{UTS}, and strain to failure, ε_f, with decreasing temperature from 293 K to 198 K and 77 K. *b*. Fracture toughness tests on side-grooved C(T) specimens show an increasing fracture resistance with crack extension; K_{JIc} values are 208 MPa$\sqrt{m}$, 265 MPa$\sqrt{m}$ and 273 MPa$\sqrt{m}$ at 293 K, 198 K, and 77 K, respectively. *c*. Stereo microscopy images show a pronounced stretch-zone between the pre-crack and the fully ductile fracture region. *d*. Fracture surfaces show ductile dimpled fracture and chromium-rich particles that act as void initiation sides.

Adapted from Gludovatz et al. (*2*).

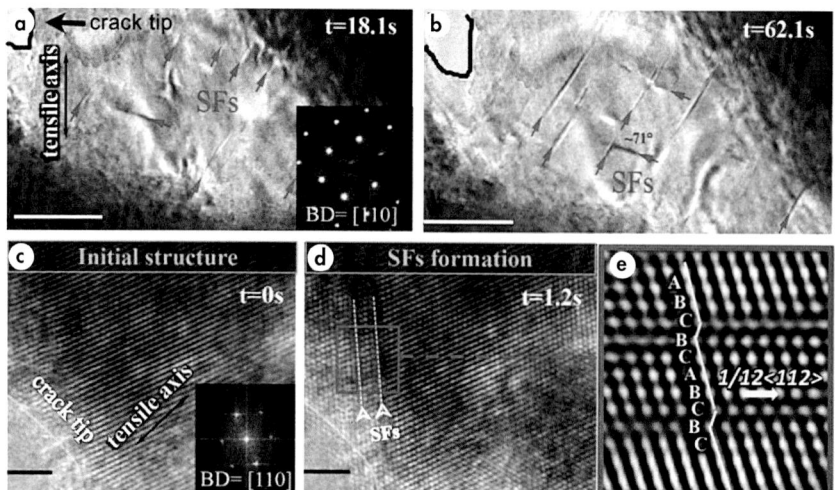

FIG. 3. — Partial dislocation activity and stacking fault formation. *a, b*. Bright-field TEM images that show the formation of stacking faults (indicated by the arrows) at the crack tip (top left-hand corner) under in situ loading of the CrMnFeCoNi high-entropy alloy (scale bars = 50 nm). Beam direction is [110]. *c, d*. High-resolution TEM images (scale bars = 2 nm) showing the formation at the atomic scale of multiple stacking faults (SFs) at the crack tip (bottom left-hand corner). *e*. The magnified inverse fast Fourier transform image showing the atomic structure and stacking sequence of the stacking faults, surrounded by the box in panel d. After Zhang et al. (*3*).

At relatively low strains (~2%), deformation in these alloys occurs on normal fcc {111} slip systems by planar dislocation glide; both $1/2$ <110> type dislocations and stacking faults are observed with the splitting of some perfect dislocations into Shockley $1/6$ <112> type partials. At higher strains (~20%), nano-twinning becomes an additional deformation mechanism, at ambient temperatures in the stronger CrCoNi alloy and at lower temperatures, i.e., ~77 K, in CrMnFeCoNi. We have argued that since the combination of dislocation slip and twinning provides a steady source of strain hardening to inhibit necking, ductility is enhanced together with strength to give excellent toughness. However, the precise nano-scale mechanisms have remained uncertain. In this presentation the origins of such remarkable damage-tolerant properties are examined for these CrCoNi-based entropy alloys, where the strength and ductility are both progressively increased at cryogenic temperatures, using in situ deformation and fracture experiments in transmission and scanning transmission electron microscopes.

II. Nano-Scale Mechanisms of Deformation and Toughening

As an example of the remarkable damage-tolerant properties of these alloys, reproduced in Fig. 2 are the uniaxial tensile properties and $J-R$ resistance curves, together with the

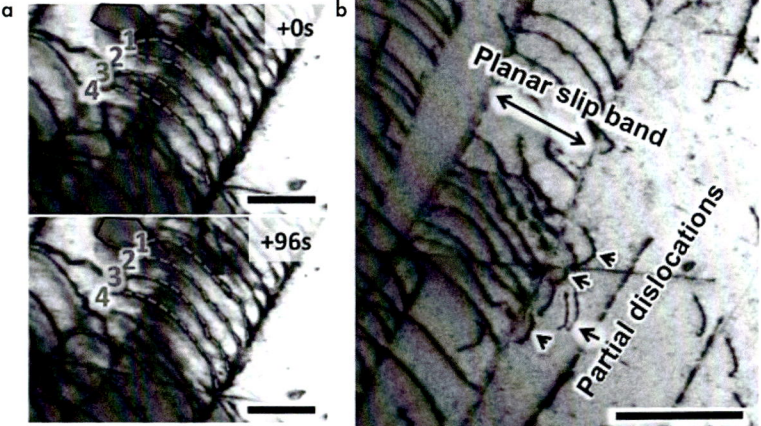

FIG. 4. — Slow planar slip of perfect dislocations. *a.* TEM images represent the dynamic process of the planar slip of undissociated $1/2<110>$ type dislocations in the CrMnFeCoNi alloy; their motion was observed to be slow and quite difficult, in contrast to the easy motion of the $1/6<112>$ type partials (shown in Fig. 3). Scale bars = 200 nm. *b.* A bright field TEM image showing the blocking of partial dislocations by the localised band of planar slip (scale bar = 500 nm). Partial dislocations move fast but can be abruptly arrested at the localised bands of planar slip containing arrays of many closely-packed perfect dislocations. The strong interaction between them provides a significant hardening effect.

After Zhang et al. (*3*).

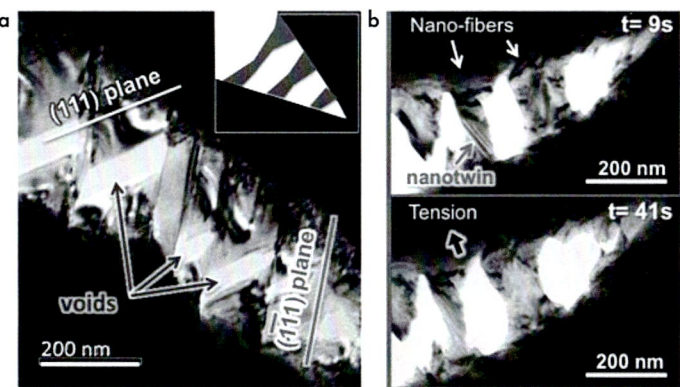

FIG. 5. — Crack bridging via near-tip twinned nanobridges in CrMnFeCoNi. *a.* Bright-field TEM image of a growing crack during in situ straining, showing submicron voids formed at the intersection of two {111} slip systems; the crack tip is located ~500 nm away from the right-lower corner. Inset is a schematic of the crack-tip structure. *b.* Two TEM images show the tensile loading of nano 'fibres' that bridge the crack in the near-tip region; nanotwins form in some 'fibres', which enhances their ductility.

After Zhang et al. (*3*).

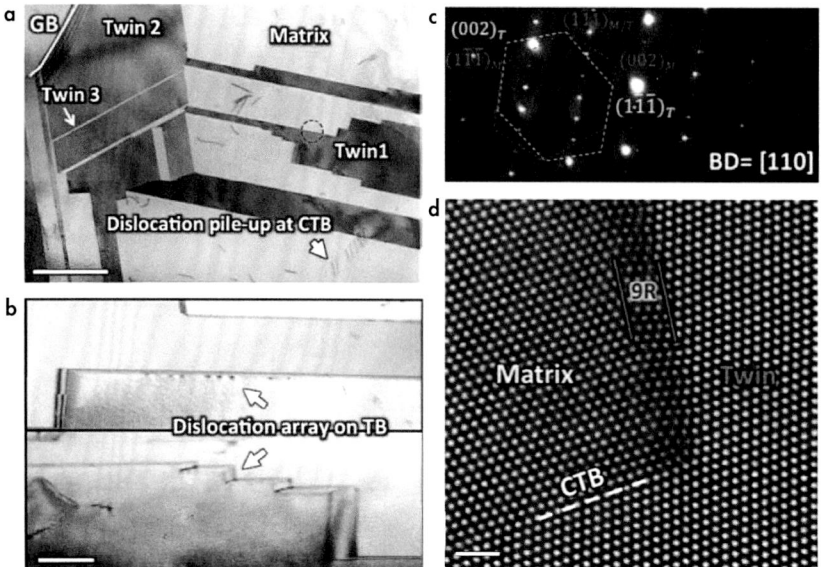

FIG. 6. — TEM of twin structures in the CrCoNi alloy. *a.* Bright-field TEM image showing the hierarchical twinning architecture in a grain of the CrCoNi alloy. A grain boundary is marked by the yellow line near the top-left corner, and multiple twinning systems are labelled. Scale bar = 1 µm. *b.* Low magnification bright-field TEM image showing dislocation arrays on the twin boundary. Scale bar = 500 nm. *c.* Selected-area electron diffraction pattern along <110> beam direction from the region on the coherent twin boundary (CTB) circled in showing extra spots which belong to the twin. *d.* High-angle annular dark field scanning TEM image showing the structure of a CTB and an incoherent twin boundary (ITB) which contains a 9R structure; the image was taken from an intersection of CTB and ITB of twin 1 in panel a. Scale bar = 500 µm.

After Zhang et al. (*4*).

highly ductile fracture surfaces, for the CrCoNi alloy at temperatures between ambient and liquid nitrogen temperatures (*2*). With tensile strengths approaching 1.4 GPa at 77 K, the measured K_{JIc} fracture toughness values rise to 270 MPa$\sqrt{\text{m}}$ at this temperature, with the crack growth toughness at a breathtaking 450 MPa$\sqrt{\text{m}}$ — all are perfectly 'valid' numbers in terms of the ASTM Standard 1820 for fracture toughness testing.

Similar results, although not quite as good, are found for the CrMnFeCoNi Cantor high-entropy alloy, which displays tensile strengths rising to 1.4 GPa with valid K_{JIc} fracture toughness values of 220 MPa$\sqrt{\text{m}}$ at 77 K (*1*).

Using principally in situ straining experiments in an aberration-corrected transmission electron microscope (TEM), we have tried to interpret the origin of these properties in terms of the underlying atomistic to micro-scale mechanisms. By imaging as close as within a few hundred nanometers of the crack tip, we were able to identify multiple deformation mechanisms associated with these alloys' high friction stress yet low stacking fault energy (SFE), that are activated at different stages of deformation and act synergistically to contribute to the toughness (*3,4*). In the CrMnFeCoNi alloy at room temperature (*3*), deformability is afforded by the easy motion of Shockley partial dislocations with the corresponding formation of stacking faults (Fig. 3). However, as the applied stress increases, perfect dislocations start to move but their slow motion is restricted to closely-packed arrays within localised planar bands (Fig. 4); these bands additionally act as strong barriers for partial dislocation motion, which creates an outstanding strengthening effect. Strengthening is also caused by strange sessile parallelepiped volume defects formed by the interaction of partials slipping on different planes that block the motion of other dislocations. Finally, in the later stages of deformation, nano-scale bridges are created that span the crack-tip region and deform by nano-twinning; as these bridges carry load that would be otherwise used to propagate the crack, they provide a potent extrinsic means to inhibit crack advance (Fig. 5).

Similar TEM studies (*4*) on CrCoNi medium entropy alloy show that a hierarchical twin network is established at ambient temperatures, associated with its very low SFE (see Fig. 6). This network generates substantial 3D barriers to dislocation motion, contributing to high strength and marked strain hardening. Simultaneously though, this twin network provides multiple pathways for the easy motion of both partial and full dislocations, which can travel directly along the twin boundaries, as shown in the TEM images in Fig. 7. This provides a source of significant plastic deformation which provides for ductility. As a result, again through a duality of deformation mechanisms, these CrCoNi-based entropy alloys can generate high strength coupled with high ductility and

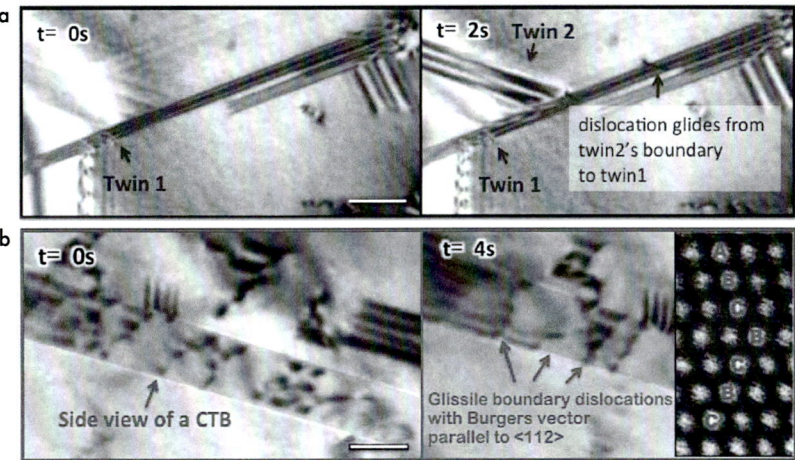

FIG. 7. — In situ imaging shows the dislocation and twin sequences during deformation. *a*. TEM image sequence showing the formation of the twinning architecture and the dynamic process of dislocations gliding from one twin boundary to another. Scale bar = 200 nm. *b*. TEM images showing the glide of paired partial dislocations on a CTB. The inset on the right shows the change of stacking at the twin boundary from fcc to hcp due to the glide of a partial dislocation. Scale bar = 200 nm.

After Zhang et al. (*4*).

continuous strain hardening. This presents the perfect ingredients for their exceptionally high fracture toughness and outstanding damage-tolerance (*1*, *2*).

III. Conclusions

Nonlinear-elastic fracture mechanics testing, coupled with in situ straining studies in aberration-corrected transmission electron microscopes, show that CrCoNi-based medium- and high-entropy alloys can display remarkable damage-tolerance, especially at cryogenic temperatures, in terms of their unprecedented combinations of strength, tensile ductility and fracture toughness. Indeed, it is believed that these multiple-element metallic alloys are some of the toughest materials reported to date. The fundamental, atomic-scale, origin of these properties appears to be associated with a synergy of dislocation mechanisms, coupled with extensive deformation nano-scale twinning, processes that are triggered at different stages of deformation to both inhibit the motion of certain dislocations and enhance the mobility of others, thereby promoting the simultaneous development of the strength and ductility that is the basis of their exceptional toughness.

Acknowledgements

This research was supported by the US Department of Energy, Office of Science, Basic Energy Sciences, Materials Sciences and Engineering Division under contract no. DE-AC02-05CH11231 to the Lawrence Berkeley National Laboratory. Thanks are due to Professor Easo George at Oak Ridge National Laboratory for processing the alloys and his insight into their deformation behaviour, Professor Bernd Gludovatz, now at UNSW in Sydney, Australia, who as my post-doc at Berkeley did all the mechanical testing, and Professor Qian Yu, formerly of Berkeley but now at Zhejiang University in Hangzhou, China, for her brilliant microscopy.

References

1. B. Gludovatz, A. Hohenwarter, D. Cartoor, E. H. Chang, E. P. George, and R. O. Ritchie, *Science* **345**, 1153 (2014).
2. B. Gludovatz, A. Hohenwarter, K. V. S. Thurston, et al., *Nat. Comm.* **7**, 10602 (2016). https://doi.org/10.1038/ncomms10602
3. Z. Zhang, M. M. Mao, J. Wang, et al., *Nat. Comm.* **6**, 10143 (2015). https://doi.org/10.1038/ncomms10143
4. Z. Zhang, H. Sheng, Z. Wang, et al., *Nat. Comm.* **8**, 14390 (2017). https://doi.org/10.1038/ncomms14390

Assessment of Non-Sharp Defects Using the Notch Failure Assessment Diagram

A. J. Horn

ABSTRACT.—The failure assessment diagram (FAD) is used to assess structures containing defects, and this approach inherently assumes that the defects being assessed are sharp. Whereas this may be appropriate for fatigue cracks, the assumption is pessimistic for flaws that do not have sharp tips such as manufacturing defects, porosity, lack-of-fusion, mechanical damage, or certain design features such as crevices. Over the last 25 years, several notch failure assessment diagram (NFAD) approaches have been proposed to assess non-sharp defects. The aim of this presentation is to provide an overview of the technical basis behind the NFAD approach, and to highlight a unique technical challenge arising from the use of notched compact tension (CT) specimens. These are affected not only by the in-plane effect of the notch, but also by out-of-plane constraint loss which is itself enhanced by the presence of the notch radius.

I. Introduction

The structural integrity of steel components is conventionally assessed using fracture mechanics based defect assessment procedures (*1 – 3*). For a real or postulated defect, the crack driving force under the loading conditions and temperature of interest is compared with the material's fracture toughness. For very brittle materials such as ceramics or glass, toughness is typically described using the linear elastic plane strain fracture toughness K_{Ic}. Conversely, for structural steel, significant plastic deformation can occur prior failure both by brittle cleavage fracture and ductile tearing, hence structural integrity is assessed using elastic–plastic fracture mechanics which may be applied through the use of a failure assessment diagram (FAD). FAD approaches necessarily use descriptions of material toughness consistent with elastic–plastic fracture mechanics, for example the critical crack tip opening displacement (CTOD), or the critical J-integral, J_{mat}. This is often expressed in dimensions of K for convenience and referred to as K_{mat}, despite being an elastic–plastic parameter.

In the FAD, the ordinate K_r indicates the proximity to fracture. K_r is defined as K_I/K_{mat}, where K_I is the linear elastic stress intensity factor and K_{mat} is the material's elastic–plastic fracture toughness. The abscissa L_r indicates the proximity to failure by plastic collapse and is defined as P/P_L where P is the applied load and P_L is the plastic limit load. K_r and L_r are both proportional to P and a linear loading line can be plotted on the FAD. When all inputs are best-estimate values, failure is predicted at the intersection of the loading line with the failure assessment curve (FAC) which

---

Wood plc, Bridgewater Place 305, Birchwood Park, Warrington, Cheshire, UK.

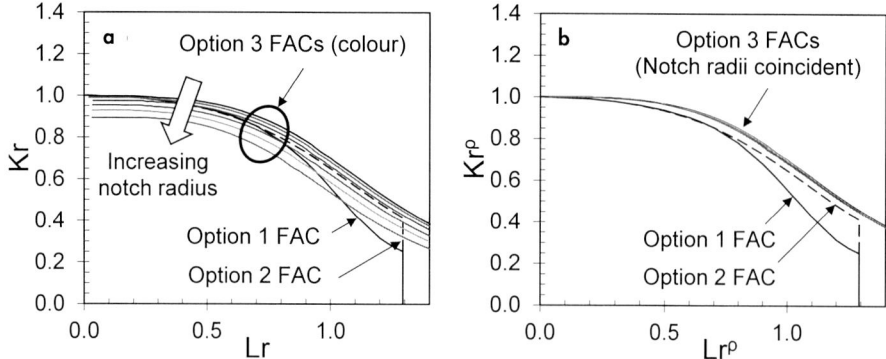

Fig. 1. — a. NFAD with axes defined as for the FAD. b. NFAD with redefined axes. After Horn and Sherry (9).

is represented by $K_r = f(L_r)$ for $L_r < L_{r(\max)}$ where $f(L_r)$ is the FAC, $L_{r(\max)}$ is the ratio of the uniaxial flow stress to the uniaxial yield stress σ_y defined at 0.2% plastic strain, and flow stress is defined as the mean of σ_y and the ultimate tensile strength (UTS). The FAC essentially acts as a conversion between linear elastic and elastic–plastic fracture mechanics, enabling K_I to be compared to K_{mat}. Three options are available for constructing the FAC using a graded approach (1, 2). The 'Option 1' curve is the simplest option: it is material and geometry independent, it is the easiest to construct requiring no geometry or tensile data (other than to define $L_{r(\max)}$), but is also the most conservative. The 'Option 2' curve is less conservative: it is independent of geometry but dependent on tensile properties and requires knowledge of the material's full stress–strain curve. Where the full stress–strain curve is unavailable an 'Approximate Option 2' can be used based on limited tensile data. The 'Option 3' curve is the most accurate and least conservative curve: it is geometry and material dependent and must be calculated from elastic–plastic finite element (FE) analysis of the cracked structure or component using the equation $f_3(L_r) = \sqrt{J_e/J}$, where J_e is the elastic component of the J-integral. To assess a steel component or structure containing a crack of depth a, K_r and L_r are calculated and an assessment point is plotted on the FAD. If the assessment point falls inside the FAC the component is safe, whereas if the assessment point falls outside the FAC the component is potentially unsafe.

The FAD approach assumes all defects to be infinitely sharp. Although this may be an appropriate assumption for fatigue cracks, it can be pessimistic for a range of other types of defects that can occur in fabricated steel structures such as porosity, mechanical damage, corrosion damage, lack of fusion, or some design features such as crevices in tube to tube-plate assemblies. Over the last 25 years, several notch failure assessment diagram (NFAD) methods have been published in the literature for assessing non-sharp defects using a modified form of the FAD (4–9). The exact form of the NFAD varies between the different approaches, but one similarity common to all NFAD approaches is the requirement to use an effective notch toughness in place of the material's fracture toughness. In some forms of the NFAD a critical linear elastic notch stress intensity factor K_{Ic}^{ρ} is used in place of the plane strain fracture toughness K_{Ic} (5, 6). For the assessment of structural steels where K_{mat} is used to describe elastic–plastic fracture toughness, NFAD approaches (which use an elastic–plastic description of effective notch toughness) can be used. In this paper, this elastic–plastic effective notch toughness is referred to as K_{mat}^{ρ}.

If the axes of the NFAD are left unchanged from those of the FAD, the 'Option 3' failure assessment curves are dependent on ρ notch-root radius as shown in Fig. 1 a. However, some NFAD approaches redefine the axes such that $K_r^{\rho} = K_I^{\rho}/K_{mat}^{\rho}$, where K_I^{ρ} is the linear elastic stress intensity factor for a notch, and $L_r^{\rho} = P/P_L^{\rho}$, where P_L^{ρ} is the plastic limit load for a notch (4, 9). In doing so the 'Option 3' FACs become independent of ρ and also lie outside the 'Option 1' and 'Option 2' curves, as shown in Fig. 1 b. This is a useful result as it means the existing FACs used in the FAD can also be used in the NFAD when axes are defined using K_r^{ρ} and L_r^{ρ}.

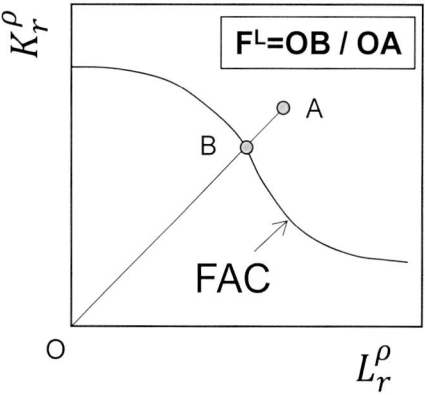

FIG. 2. — Definition of F^L.

A reserve factor on load, F^L, can be defined to provide an indication of the margins against failure. F^L is defined in Fig. 2 where A is the assessment point and B is the limiting condition. Assessment points inside the FAC have $F^L > 1$, and points outside the FAC have $F^L < 1$.

II. Linking K_{mat}^{ρ} to Notch Geometry

In order to characterise how K_{mat}^{ρ} changes in the presence of a notch, a measure of the notch geometry is required. Traditionally this has been described using either the notch-root radius ρ (4, 7, 9), or the elastic stress concentration factor k_t (5, 6).

One of the key challenges in developing any engineering method for predicting notch fracture is defining the relationship between K_{mat}^{ρ} and the parameter chosen to describe the notch geometry such as ρ or k_t. This relationship will be dependent on the parameter used to characterise the notch geometry, as shown in Fig. 3, and can be developed in a variety of ways, for example using experimental testing (7–10), application of notch fracture theories (11–16), or FE modelling incorporating a suitable local approach failure criterion capable of predicting notch fracture by cleavage (9, 17), or ductile tearing (18, 19). This paper uses the example of the Beremin model (20) to illustrate some general concepts relating to the development of engineering assessment methods for notch fracture.

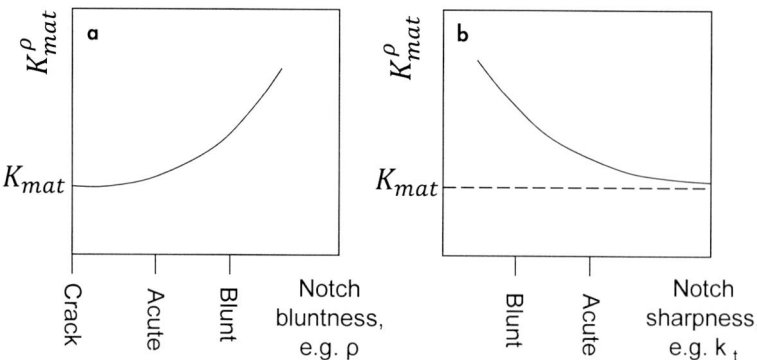

Fig. 3. — Schematic showing relationship between K^ρ_{mat} notch bluntness (a), and notch sharpness (b).

III. Toughness Scaling

The Beremin model describes the proximity to cleavage fracture by use of the scalar Weibull stress σ_w. In its simplest form the probability of cleavage fracture is described by a two-parameter Weibull distribution

$$[1] \qquad P_f = 1 - \exp\left[-\left(\frac{\sigma_w}{\sigma_u}\right)^m\right] ,$$

where σ_u and m are the Weibull parameters. The Weibull stress σ_w is calculated by integrating a weighted value of the maximum principal stress σ_1 over the plastic zone V ahead of the stress concentration

$$[2] \qquad \sigma_w = \left[\frac{1}{V_0}\int_V \sigma_1^m \, dV\right]^{1/m}$$

The constant V_0 is a reference volume required to ensure dimensional consistency and is here taken as unity. The Weibull parameters σ_u and m are calibrated by matching values calculated from the Beremin model to experimental values of cleavage K_{mat}. Reliable estimation of the Weibull parameters is only possible using experimental data of sufficient quantity that cover two different constraint levels. The method proposed by Gao et al. (*21*) provides a suitable methodology.

1. Toughness Scaling Using Notch-Root Radius

For a given material and geometry, the evolution of σ_w with loading (defined here using J for a notch, J^ρ) can be plotted, for example as shown in Fig. 4 *a* for three-point bend

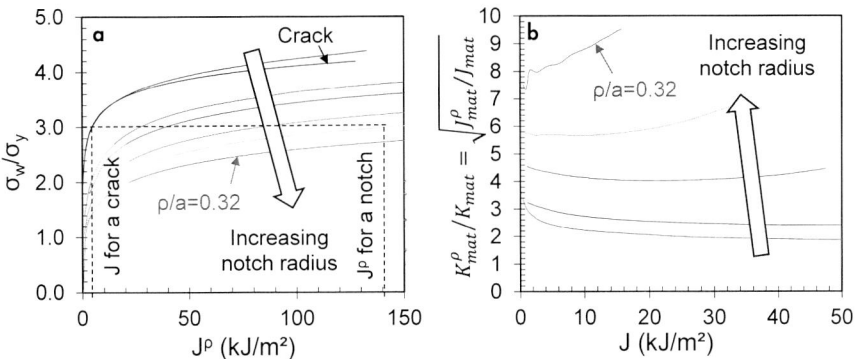

FIG. 4. — Toughness scaling: lines of constant notch radius.

specimens containing U-notches of different ρ. The Weibull stress is directly related to the probability of cleavage fracture through Eq. [1], therefore a horizontal line plotted in Fig. 4 a defines a particular cleavage fracture probability. The horizontal line can then be used to scale any value of J_{mat} for a crack to the equivalent J^ρ_{mat} value that a notch would have for the same probability of cleavage fracture. The curves in Fig. 4 a can therefore be used to define the toughness benefit J^ρ_{mat}/J_{mat} (or K^ρ_{mat}/K_{mat}) for any value of J (or K_J), as shown in Fig. 4 b.

The advantage of this toughness scaling approach is that because the necessary parameters (in this case σ_w and J^ρ) can be generated from 3D FE models, the approach can be used to convert from any cracked geometry to any notched geometry. This type of toughness scaling approach was used by Brown et al. (22) to develop the latest guidance for assessing gouges in pressure vessels in API 5791/ASME FFS1 (3). Weibull stress analysis was used to study cleavage behaviour, and the Bao–Wierzbicki ductile failure model (23) was used to undertake toughness scaling on the upper shelf. The geometry addressed by Brown et al. (22) is that of a longitudinally-oriented gouge on the external surface of a pressurised cylinder, with a weld residual stress which varies linearly through the thickness of the cylinder from tension at the outer surface to compressive at the inner surface. The material properties used were those of a commonly used pressure vessel steel. For specific industrial problems such as this, where the range of material, loading and geometrical combinations is limited, the toughness scaling approach described above is very convenient to use. Confidence in the approach can be demonstrated by validation of the FE models and of the local failure criterion which must be shown to be applicable to notches.

However, for practical reasons this toughness scaling approach is less convenient for developing more general guidance capable of assessing a range of geometries or materials. Detailed elastic–plastic FE models must be analysed and relevant local approach failure criteria invoked for every required combination of material, applied loading, residual stress distribution, component geometry, component thickness, defect depth, defect tip radius, defect orientation, and whether the defect is embedded, fully through-thickness, or surface breaking on the inside or outside of the component. For anything other than a specific industrial problem, the number of FE analyses that would need to be generated would be substantial to cater for all possible combinations of material properties, component geometries, defect geometries and loading scenarios that a structural integrity engineer may need to consider.

Furthermore, Fig. 4 *b* shows that this form of toughness scaling also results in toughness benefits K^ρ_{mat}/K_{mat} that are dependent on load, failure probability, and hence K_{mat} itself. This has two important implications. Firstly, it becomes even less practical to develop general expressions for K^ρ_{mat}/K_{mat} when the ratio varies with applied load in addition to the factors listed above. Secondly, different guidance would be required depending on the K_{mat} bound used as an input to the FAD or NFAD: when assessing a real or postulated defect, a suitable lower bound toughness value must be selected. What constitutes a *suitable* lower bound is not necessarily trivial and in some industries it is related to the required reliability of the component being assessed. For example, in the nuclear industry, it must be demonstrated that very high integrity components with severe consequences of failure have an extremely low probability of failure, and this is achieved using a 2.5% lower bound K_{mat} value in the FAD assessment, together with other conservative inputs and onerous target reserve factors to ensure adequate reliability. Other toughness bounds may also be used in certain circumstances, for example when assessing very rare but extreme loading conditions which have a low probability of occurrence. In non-nuclear applications where the consequences of failure are less severe, BS7910 can be used which recommends the use of a 20% bound on K_{mat} where failure results in the loss of function of the component and no direct loss of life, or a 5% bound where failure would result in loss of life (*2*). In contrast, when using the FAD for research purposes, it is useful to use the median fracture toughness as this should result in 50% of the test data being inside and 50% outside the FAC (with all other inputs being best-estimate values).

The FAD is clearly a very flexible assessment tool which is used in a wide variety of industrial sectors and can cater for differing requirements on the chosen toughness bound. For a generalised NFAD approach to be equally flexible across different industries with differing reliability requirements, it is therefore desirable to use a different form of toughness scaling approach that results in ratios of K^ρ_{mat}/K_{mat} that are independent of load, independent of failure probability, and therefore also independent of the toughness bound used to define K_{mat}.

One method of doing this is to introduce an appropriate load dependency into the description of the notch geometry, i.e. by using a new load-dependent parameter in place of ρ. One convenient load-dependent parameter that describes the notch geometry is the component of the elastic notch-tip stress acting in a direction normal to the plane of the notch, referred to in this paper as σ_N.

2. Toughness Scaling Using Elastic Notch-Tip Stress

The elastic notch-tip stress, σ_N, scales with load: a given value of σ_N could correspond to an acute notch under a low load, or a blunter notch subject to a higher load. A

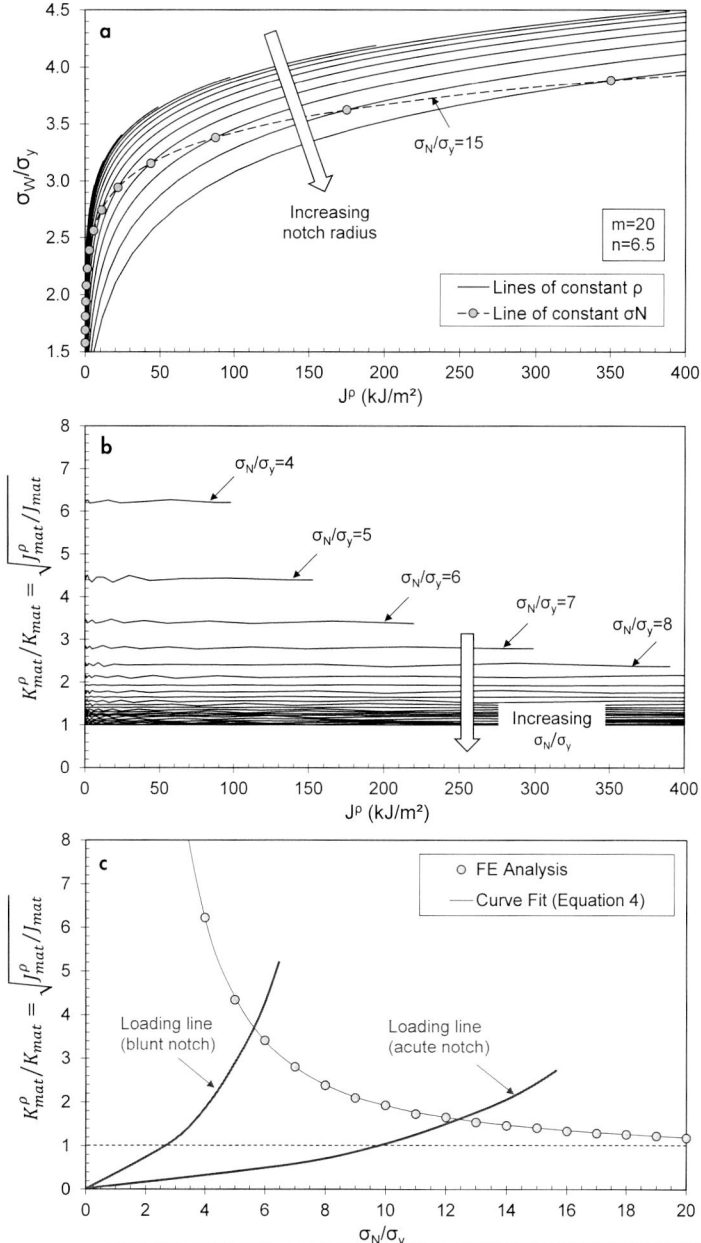

Fig. 5. — Toughness scaling using boundary layer models. *a*. Converting lines of constant ρ to lines of constant σ_N/σ_y. *b*. Toughness benefit for constant σ_N/σ_y. *c*. Toughness benefit *vs* σ_N/σ_y. After Horn and Sherry (*9*).

convenient expression for σ_N was derived by Shin (24) based on the Creager–Paris elastic stress distribution (25) ahead of a slender notch in a uniform stress field

[3] $$\sigma_N = \sigma(1 + 2Y\sqrt{a/\rho}) ,$$

where σ is the applied tensile stress remote from the notch, a is the notch depth and Y is a geometry factor used to define K.

Fig. 5 a shows the evolution of σ_w with J^ρ for notches in an infinite solid obtained using boundary layer FE analyses reported by Horn and Sherry (9). Each solid line is an output from a different boundary layer analysis and is therefore a loading line of constant ρ and analogous to the lines plotted in Fig. 4 a. For each FE model, the point on the loading curve at which $\sigma_N/\sigma_y = 15$ has been marked by a solid grey circle. Joining these points together results in the dashed line which is a locus of constant σ_N/σ_y. This process was repeated for different σ_N/σ_y values and the toughness scaling process was then performed on the lines of constant σ_N/σ_y instead of the lines of constant ρ. The resulting toughness benefit shown in Fig. 5 b is independent of J^ρ (note how this differs from Fig. 4 b). For the reasons discussed above, this is a particularly useful result: values of K_{mat}^ρ/K_{mat} for constant σ_N/σ_y are independent of load, independent of σ_w and therefore independent of failure probability and toughness bound. Values of K_{mat}^ρ/K_{mat} for constant ρ (i.e. a specific geometry) *are* dependent on load as shown by Fig. 4 b. When describing K_{mat}^ρ/K_{mat} for constant σ_N/σ_y the load dependence comes through the parameter σ_N, which can be obtained easily from the applied load using Eq. [3]. The desired failure probability or toughness bound comes through the selection of the appropriate K_{mat} value.

Using the results shown in Fig. 5 b, it is then possible to plot a curve of toughness benefit against σ_N (Fig. 5 c). A loading line for the same material may be plotted on the diagram for any component and notch geometry of interest, with σ_N being calculated via Eq. [3]. Failure is predicted by the intersection of the loading line with the failure locus. For blunt notches the loading curve rises steeply and failure is predicted at high K_{mat}^ρ values, and for acute notches the loading curve is less steep and failure is predicted at lower K_{mat}^ρ values. As $\rho \to 0$, the gradient of the loading line approaches the horizontal and $K_{mat}^\rho \to K_{mat}$.

The following empirical power law expression was found to describe the increase in effective cleavage toughness with increasing notch radius in Horn and Sherry (9), and preliminary work by Kim et al. (26) indicates that the same expression may be used to describe the increase in ductile tearing initiation toughness with notch radius

[4] $$\frac{K_{mat}^\rho}{K_{mat}} = 1 + \gamma \left[\frac{\sigma_N}{\sigma_u}\right]^{-l} ,$$

where γ and l are non-dimensional material properties that describe the sensitivity of material toughness to the notch-root radius. Depending on failure mechanism, values of and that define the failure locus in Fig. 5 c can be obtained in one of three ways:

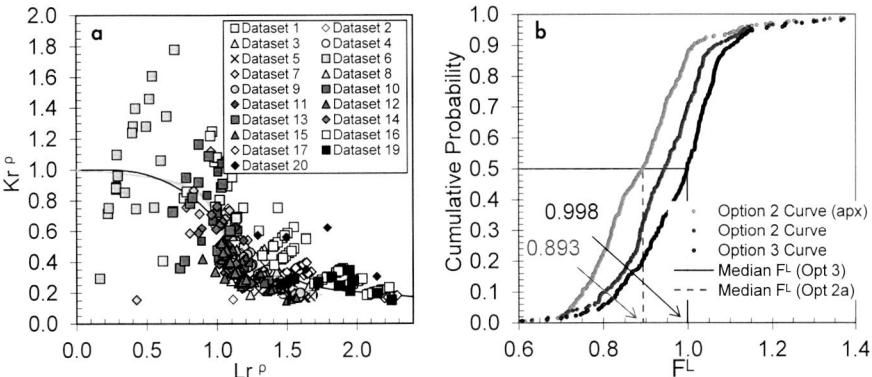

FIG. 6. — NFAD validation (K_{mat} defined at $P_f = 50\%$).
After Horn et al. (29).

(i) micromechanical modelling, for example as shown in Fig. 5 a; (ii) use of the lookup table in Horn and Sherry (9) based on a knowledge of Weibull modulus m and strain hardening exponent n; or (iii) by curve fitting to test data. Curve fitting is discussed in the following section.

Once the material parameters γ and l have been obtained for the material of interest, $K^\rho_{\text{mat}}/K_{\text{mat}}$ can be calculated for any geometry and any level of loading via Eq. [4]. The parameter σ_N can be obtained from the applied load using Eq. [3]. Y is a geometry term used to define K_I and once K_I has been calculated using the standard FAD approach, Y is simply calculated as $Y = K_I/(\sigma\sqrt{\pi a})$.

IV. NFAD Validation

The NFAD approach developed by Horn and Sherry (9) was validated for one steel at one temperature in Horn and Sherry (27), and for 20 datasets covering seven steels tested over a wide range of temperatures — a total of over 500 individual data points — by Horn et al. (28, 29). Defining K_{mat} as the median value (i.e. $P_f = 50\%$), around half of the failure assessments points lie outside the most accurate 'Option 3' failure curve (i.e. $F^L < 1$), and half inside (i.e. $F^L > 1$), as shown in Fig. 6 a. Fig. 6 b is a cumulative probability plot which shows that the median F^L value is 0.998 when defined using the most accurate Option 3 failure assessment curve, confirming a very nearly even split between assessment points falling inside and outside the failure curve. F^L values are slightly lower when calculated using the more conservative 'Option 2' and approximate 'Option 2' curves, as would be expected.

Overall, this validation work provides confidence in the NFAD assessment methodology for non-sharp defects for several materials tested at different temperatures. However, this confidence is limited to the occasions where γ and l are obtained by curve fitting

TABLE
Summary of size and geometry effects in cracked and notched specimens and components.

In-plane, or out-of-plane	Type of effect	Cause of effect	Analysis approach
In-plane	Mechanical	Notch radius	σ_N
		Geometry and loading	T-stress (contained yielding)
			Q-parameter (extended yielding)
Out-of-plane	Mechanical	Plane stress *vs* plane strain	No standardised approach
	Microstructural*	Crack-front length*	Weakest link statistics*

* Cleavage fracture only.

to test data from specimens that are the same size as those used for assessments. Validation work carried out to date does not address the much more likely application of the NFAD where the test specimens will be of a different size and geometry to the structure or component being assessed. This raises the question: can values of K_{mat}^ρ measured on test specimens be used in an NFAD assessment of a structure with a different size and shape?

V. Size and Geometry Effects

Several different types of size and geometry effects can occur in notched and pre-cracked test specimens and structures. NFAD approaches in the literature typically only address the in-plane mechanical (constraint loss) effect of the notch radius on K_{mat}^ρ. However, there are other forms of mechanical (constraint loss) and microstructural effects that can affect values in specimens and structures containing cracks. These also have potential to affect values of K_{mat}^ρ measured using notched test specimens.

The table provides a summary of the different types of size and geometry effects that can occur in notched and pre-cracked specimens and structures. The top row covers typical NFAD approaches which account for the in-plane constraint loss caused by the notch radius. The remaining rows in the table show other size and geometry effects which can occur. The T-stress and Q-parameter approaches are often used to describe in-plane constraint loss in pre-cracked specimens. This form of constraint loss occurs when the crack-tip stress fields are less constrained and cause the fracture toughness K_{mat} to become geometry dependent. Typical conditions that cause this form of constraint loss are: geometrical (e.g. shallow cracks where the crack-tip stress fields interact with the free surface just behind the crack tip); loading type (e.g. tension rather than bending which promotes interaction of the crack-tip stress fields with free surfaces early in the loading), or loading extent (e.g. at high loads where the extent of the crack-tip plastic zone becomes large and interacts with free surfaces). Fracture toughness testing standards are designed to ensure that the crack-tip plastic zone is highly constrained, thereby ensuring the measured K_{mat} is a geometry-independent value. However, a common problem for structural integrity assessment is that of shallow cracks or structures loaded in tension, where the constraint-dependent toughness K_{mat}^c of the structure may be significantly higher than the high-constraint value K_{mat} measured in tests.

Out-of-plane constraint loss can also occur in pre-cracked specimens, however testing standards have minimum thickness requirements to ensure that plane strain conditions dominate and the out-of-plane constraint loss is sufficiently small that the K_{mat} value measured corresponds to plane strain conditions.

The last row of the table refers to an out-of-plane microstructural effect which is relevant only for cleavage fracture. For thick specimens, a longer crack-front length and hence larger plastic zone volume increases the probability of sampling a microstructural feature capable of triggering cleavage fracture compared with a thin specimen. A thickness correction to account for this effect is provided in the master curve standard (29).

Recent modelling work by Horn et al. (30, 31) has shown that for notched compact tension (CT) specimens, K_{mat}^ρ is a function of not only the in-plane effect of the notch radius but also out-of-plane constraint loss which itself is enhanced by the presence of the notch radius. In other words, K_{mat}^ρ is dependent on specimen thickness B in a way that standard pre-cracked CT specimens are not. This out-of-plane constraint loss can be so significant that it can increase K_{mat}^ρ more than the in-plane notch radius effect alone. The use of experimentally measured K_{mat}^ρ values in an NFAD assessment of a notched structure may therefore be non-conservative if the out-of-plane constraint loss in the notched CT specimen is more significant than in the structure. For example, a long surface-breaking non-sharp defect in a structure would be predominantly in plane strain, but tests on standard thickness CT specimens of the same material may exhibit much higher values of than in the structure due to the out-of-plane constraint loss in the CT specimens. Using K_{mat}^ρ values from these CT tests to assess failure in a structure where plane strain conditions dominate would result in a non-conservative NFAD assessment.

Investigations by Horn et al. (30, 31) showed that doubling the thickness B of the standard CT specimen resulted in plane-strain conditions being sufficiently dominant to eliminate out-of-plane constraint loss in the notched CT specimen. The work also indicated that the other size effects listed in Table 1 are not noticeably enhanced by the presence of the notch radius, however as only one material has been considered, further work is required to determine whether these conclusions hold more widely.

VII. Discussion

The key advantage of geometry-specific toughness scaling approaches, such as that used to develop the guidance in API 5791/ASME FFS1, is that the stresses and strains calculated using elastic–plastic FE models of the notched component of interest inherently capture all forms of constraint loss via the evolution of the crack or notch-tip plastic zone. The use of the volume-based Weibull stress approach also captures microstructural crack (or notch) front length effects. There is therefore no need to fully understand nor describe the effects in the table separately, as they are already inherently accounted for by the FE models of the components of interest.

The key challenge for developing a more generalised method for assessing notch fracture in steels is the opposite of this: the effect of the notch radius on all effects listed in the table must be fully understood and, where effects are significant, suitably characterised to take full advantage of the presence of a non-sharp defect. The NFAD assessment method (9) currently being considered for inclusion in the R6 guidance document (1) is capable only of accounting for the top row in the table. Guidance is provided for minimising the out-of-plane constraint loss in notched CT specimens to ensure that measured values of K_{mat}^{ρ} correspond to plane strain conditions. So although safeguards have been put in place to ensure the NFAD predictions remain conservative, there is potential to reduce conservatism of the assessment methodology by developing an approach for predicting the extent to which out-of-plane constraint loss is enhanced by the presence of a notch.

One of the other key challenges for applying NFAD approaches is measuring the notch-root radius ρ. Where a sufficient quantity of material is available, or where it is possible to remove samples from the structure, metallographic sectioning can be used to provide a physical measure of the notch-root radius. For certain types of non-sharp defects it may also be possible through multiple metallographic sectioning to build up a distribution of ρ values from which an appropriate lower bound could be defined. Efforts are also being made to develop non-destructive examination (NDE) techniques capable of quantifying the root radius of volumetric defects (e.g. 32, 33).

A more fundamental challenge to gaining regulatory approval for using an NFAD in a structural integrity safety case will be providing sufficient confidence that use of the NFAD is appropriate for each case. For example, it will be necessary to provide confidence that the defect being assessed is truly non-sharp. This could be undermined if there is a crack at the tip of the non-sharp defect, for example if a sub-critical crack growth mechanism such as fatigue has initiated a sharp crack. For non-sharp defects detected using NDE, confidence may need to be provided that the presence of the non-sharp defect has not masked the presence of any nearby sharp cracks. If a real or postulated sharp crack at the tip of a blunt notch is to be assessed, this could be achieved using traditional fracture mechanics of a shallow crack in a stress concentration, and potentially taking advantage of crack-tip constraint loss.

VIII. Conclusions

For toughness scaling methods based on ρ.

(1.) The constraint loss effect that the component's size and geometry has on K_{mat}^{ρ} is inherently accounted for by the use of elastic–plastic FE models of the components of interest.

(2.) It is therefore not necessary to understand or describe the effect that the different types of constraint loss have on K^{ρ}_{mat}.

(3.) The approaches are limited to the specific component geometries and loading conditions modelled using FE analysis. To extend the applicability of each approach, detailed elastic–plastic FE models must be analysed and local approach failure criteria invoked for each required combination of material properties, component geometries, defect geometries and loading scenarios. This limits the practical application for developing general guidance suitable for a wide range of geometries.

(4.) The toughness benefit $K^{\rho}_{\text{mat}}/K_{\text{mat}}$ for a given ρ is dependent on load and hence dependent on K_{mat}, which means $K^{\rho}_{\text{mat}}/K_{\text{ma}}$ also depends on the toughness bound used in the FAD.

For toughness scaling methods based on σ_N.

(1.) The constraint loss effect that the component's size and geometry has on K^{ρ}_{mat} is not accounted for; only the in-plane constraint loss caused by the presence of the notch radius is accounted for.

(2.) It is therefore necessary to understand, and where appropriate characterise, the size and geometry effects in the table to take full advantage of the presence of a non-sharp defect.

(3.) For notched CT specimens, K^{ρ}_{mat} is a function of not only the in-plane effect of the notch radius but also out-of-plane constraint loss which itself is enhanced by the presence of the notch. In other words, K^{ρ}_{mat} is dependent on specimen thickness B in a way that standard pre-cracked CT specimens are not.

(4.) These toughness scaling approaches can be applied to a wide range of geometries: they do not require FE analysis of each geometry, only K and P_L solutions which are standard parameters used in a FAD assessment.

(5.) The toughness benefit $K^{\rho}_{\text{mat}}/K_{\text{mat}}$ for a given σ_N is independent of load, independent of failure probability, independent of K_{mat}, and independent of the toughness bound used in the FAD.

Acknowledgements

The author wishes to acknowledge Tata Steel, EDF Energy and the UK's National Nuclear Laboratory (NNL) for funding the aspects of notch fracture discussed in this paper. The author also wishes to express deep gratitude for valuable discussions with Professor Andrew Sherry and Mr. Adam Bannister over the many years that this work was undertaken.

References

1. EDF, "R6—Assessment of the Integrity of Structures Containing Defects, Including Amendments 1–11", Revision 4. EDF Energy Nuclear Generation, Gloucester, UK, 2015.
2. BSI, "BS7910—Guide on Methods for Assessing the Integrity of Flaws in Metallic Structures." British Standards Institution, London, UK, 2013.
3. API, "API 579-1/ASME FFS-1, Fitness-for-Service." American Petroleum Institute, Washington, DC, USA, 2016.

4. W. Q. Wang, A. J. Li, P. N. Li, and D. Y Ju, *Int. J. Pres. Ves. Pip.* **60,** 1 (1994).
5. E. Smith, *Int. J. Pres. Ves. Pip.* **76,** 799 (1999).
6. Y. G. Matvienko, *Int. J. Fract.* **124,** 107 (2003).
7. S. Cicero, F. Gutierrez-Solana, and A. J. Horn, *Eng. Fail. Anal.* **16,** 2450 (2009).
8. S. Cicero, V. Madrazo, T. Garcia, J. Cuervo, and E. Ruiz, *Eng. Fail. Anal.* **29,** 108 (2013).
9. A. J. Horn and A. H. Sherry, *Int. J. Pres. Ves. Pip.* **89,** 137 (2012).
10. I. Milne, G. G. Chell, and P. J. Worthington, *Mater. Sci. Eng.* **40,** 145 (1979).
11. F. J. Gomez and M. Elices, *Int. J. Fract.* **127,** 239 (2004).
12. P. Livieri, *Eng. Fract. Mech.* **75,** 1779 (2008).
13. D. Taylor, P. Cornetti, and N. Pugno, *Eng. Fract. Mech.* **72,** 1021 (2005).
14. D. Taylor, *Eng. Fract. Anal.* **18,** 543 (2011).
15. E. Barati and F. Berto, *Procedia Eng.* **10,** 807 (2011).
16. L. Susmel and D. Taylor, *Eng. Fract. Mech.* **75,** 4410 (2008).
17. A. J. Horn and A. H. Sherry, *Int. J. Pres. Ves. Pip.* **87,** 670 (2010).
18. G. W. Brown, L. Parietti, B. Rose, and T. L. Anderson, *in* "Proceedings" American Society of Mechanical Engineers (ASME) 2016 Pressure Vessels and Piping Conference, 17th–21st July 2016, Vancouver, British Columbia, Canada. Volume 1A: Codes and Standards, Paper No. PVP2016-63756.
19. J. J. Han, N. Larrosa, Y. J. Kima, and R. A. Ainsworth, *Int. J. Pres. Ves. Pip.* **146,** 39 (2016).
20. F. M. Beremin, *Metall. Trans. A* **14A,** 2277 (1983).
21. X. Gao, C. Ruggieri, and R. H. Dodds Jr, *Int. J. Fract.* **92,** 175 (1998).
22. op. cit (*18*).
23. Y. Bao, *Eng. Fract. Mech.* **72,** 505 (2004).
24. C. S. Shin, *Int. J. Fract.* **8,** 235 (1986).
25. M. Creager and P. C. Paris, *Int. J. Fract. Mech.* **3,** 247 (1967).
26. J. S. Kim, N. O. Larrosa, A. J. Horn, Y. J. Kim and R. A. Ainsworth, *Eng. Fract. Mech.* **188,** 250 (2018).
27. A. J. Horn and A. H. Sherry, *Int. J. Pres. Ves. Pip.* **89,** 151 (2012).
28. A. J. Horn, S. Cicero, A. Bannister and P. J. Budden, *in* "Proceedings" American Society of Mechanical Engineers (ASME) 2016 Pressure Vessels and Piping Conference, 17th–21st July 2016, Vancouver, British Columbia, Canada. Volume 6B: Materials and Fabrication, Paper No. PVP2016-63537.
29. ASTM, "E1921-11a—Standard Test Method for Determination of Reference Temperature, T_0, for Ferritic Steels in the Transition Region." ASTM International, West Conshohocken, PA, USA, 2011.
30. A. J. Horn, S. Cicero, A. Bannister, and P. J. Budden, *in* "Proceedings" American Society of Mechanical Engineers (ASME) 2017 Pressure Vessels and Piping Conference, 16th–20th July 2017, Waikoloa, Hawaii, USA. Volume 6B: Materials and Fabrication, Paper No. PVP2017-66095.
30. A. J. Horn, A. H. Sherry, and P. J. Budden, *Int. J. Pres. Ves. Pip.* **154,** 40 (2017).
31. ASTM, "E1921-11a—Standard Test Method for Determination of Reference Temperature, T_0, for Ferritic Steels in the Transition Region." ASTM International, West Conshohocken, PA, USA, 2011.
32. J. Allwright, O. Gilmour, F. Hagglund, et al., "Development of Inspection Procedures for Non-Sharp Defects in Steel," TWI Report 22449/01/2017 (restricted circulation). TWI, Cambridge, UK, 2017.
33. P. Bastid, S. Blackwell, Y. J. Janin, G. Wu, and I. Hadley, "Assessment Procedures for Non-Sharp (Blunt) Defects," TWI Members' Report 1085/2017 (available to TWI members). TWI, Cambridge, UK, 2017.

Creep–Fatigue Crack Initiation in Weldments

M. J. Chevalier and D. W. Dean

ABSTRACT.—Weldments often exhibit poorer performance than homogeneous structures for several reasons, including the introduction of small defects, the deleterious geometry due to the presence of weld toes or undercut due to weld dressing, degradation of material properties in the heat-affected zone (HAZ) and introduction of material mismatch between weld and parent materials.

The "Assessment Procedure for the High Temperature Response of Structures" (R5) has a dedicated assessment route for the treatment of creep–fatigue crack initiation in weldments. A previous R5 approach utilised a fatigue strength reduction factor (FSRF) which affected both fatigue and creep damage calculations. The current methodology splits the FRSF into two parts: (*i*) a weldment strain enhancement factor (WSEF), which affects both fatigue and creep damage; and (*ii*) a weld endurance reduction (WER), which only affects the fatigue damage. This change in methodology was based on a physical understanding of what impact different factors have on weldment endurance. The net effect of this change in methodology is that fatigue damage calculations are generally unaffected and calculated creep damage values are reduced.

This paper describes the current and previous R5 Volume 2/3 procedures for assessing creep–fatigue crack initiation in weldments and discusses some of the more significant findings from the current TAGSI Subgroup NT-26 review on this subject.

I. Introduction

It is often the case that weldments are the structural integrity feature most prone to creep and creep–fatigue crack initiation. This has been borne out by operational experience in the EDF Energy fleet of advanced gas-cooled reactors (AGRs), where the vast majority of structural integrity challenges have been associated with welded features (*1*).

Weldments experience poorer performance than homogeneous structures for several reasons, including:
- the introduction of small defects, notably in weld metal or at the weld toe (e.g. inclusions and liquation cracking)
- the deleterious geometry due to the presence of weld toes or undercut due to weld dressing
- degradation of material properties in the heat-affected zone (HAZ)
- in cases of dissimilar metal welds, diffusion across the fusion boundary can cause local changes in material properties

---

Assessment Technology Group, Structural Integrity Branch, Nuclear Generation, EDF Energy, UK.

- the introduction of residual stresses (in cases of non-heat treated weldments)
- the introduction of material mismatch (in both plastic and creep properties) between weld, HAZ, and parent materials

The "R5—Assessment Procedure for the High Temperature Response of Structures" (*2*), has a number of dedicated assessment routes for the treatment of weldments, discussed further in Section II. This paper focusses on the creep–fatigue crack initiation assessment methodology in R5 Volume 2/3, "Appendix A4". Prior to 2014 (*3*), a single fatigue strength reduction factor (FSRF) was used in R5 to account for the impact of defects, deleterious geometry, material degradation, and material mismatch in the weldment and this factor affected both fatigue and creep damage calculations. In R5, Issue 3, Revision 2 (*2*), the methodology was updated, splitting the single FRSF into two parts: (*i*) a weldment strain enhancement factor (WSEF) which affects both fatigue and creep damage; and (*ii*) a weld endurance reduction (WER) which only affects the fatigue damage. This change in methodology was based on a physical understanding of what impact different factors have on weldment endurance. The net effect of this change in methodology is generally no change to the fatigue damage calculations and a decrease in the calculated creep damage.

This paper provides a detailed description of the treatment of creep–fatigue crack initiation in weldments within the R5 assessment procedures, in both the previous approach—use of FSRF (*3*)—and the current approach—using WSEF and WER (*2*). It then highlights some of the more significant discussion points raised by the TAGSI Subgroup NT-26, which is currently providing an independent review of the R5 Volume 2/3 "Appendix A4" assessment methodology.

II. Overview of the R5 Approach for Assessing Creep–Fatigue Initiation in Weldments

This section provides a summary of the previous weldment creep–fatigue initiation approach in Section II.1, as provided in R5, Issue 3 (*3*). The current approach (R5, Issue 3, Revision 2), is described in Section II.2 (*2*).

The R5 procedures provide an assessment of the continuing integrity of a component, where the operating lifetime might be limited by one of the following mechanisms:
- excessive plastic deformation due to a single application of a loading system
- creep rupture
- ratchetting or incremental plastic collapse due to a loading sequence
- creep deformation enhanced by cyclic load
- initiation of cracks in initially defect-free material by creep and creep–fatigue mechanisms
- the growth of cracks by creep and creep–fatigue mechanisms

R5 is an established methodology, first produced almost 30 years ago (*4*). In 2003, the former Volumes 2 and 3 of R5 were combined into a single procedure for assessing defect-free structures and the former Volumes 4 and 5 were combined into a single procedure for assessing defects in structures (*3*). In the current version of R5, Issue 3, Revision 2 (*2*) there are five volumes in total. Volume 1—"The Overview."

Volume 2/3—"Creep–Fatigue Crack Initiation Procedure for Defect-Free Structures." Volume 4/5—"Procedure for Assessing Defects Under Creep and Creep–Fatigue Loading." Volume 6—"Assessment Procedure for Dissimilar Metal Welds." Volume 7—Behaviour of Similar Weldments: Guidance for Steady Creep Loading of Ferritic Pipework Components."

Volume 1 provides an overview, which indicates the overall scope and restrictions of R5 by reference to the other volumes, provides a route for following the detailed procedures given elsewhere in R5, and compares R5 with other approaches. Volumes 6 and 7 are essentially specialised applications of the creep–fatigue damage calculations of Volume 2/3, and the creep and creep–fatigue crack growth calculations of Volume 4/5, respectively, to particular weldments and operating conditions found in AGRs.

The aims of the procedures in R5 Volume 2/3 are to estimate, by a simplified approach based on elastic stress analysis, the steady cyclic stresses and strains (the steady cyclic state) in a defect-free structure under creep–fatigue loading, and to use these parameters to estimate when creep–fatigue crack initiation will occur in the structure. The procedures are not intended to provide an estimate of the number of cycles to overall failure of a component operating at high temperature, although the crack initiation endurance is a lower bound to this. Instead, the procedures estimate the number of cycles to create a crack of a defined size. The total lifetime may then be obtained in conjunction with an assessment of the number of cycles for this crack to grow to a limiting size, by following the R5 Volume 4/5 crack growth procedures. The R5 Volume 2/3 procedures include a number of novel features such as the shakedown approach for structural assessment, the ductility exhaustion method (including multiaxial effects) for estimating creep damage, and the inclusion of size effects in fatigue damage calculations to enable assessments of thin-section components.

Conservatism in R5 Volume 2/3 assessments of creep–fatigue crack initiation is usually introduced by using best estimate steady cyclic stresses and strains (the steady cyclic state) in combination with upper bound estimates of creep and fatigue damage.

Whilst this paper focusses on weldment creep–fatigue crack initiation, this is only one such dedicated assessment route in R5 for the treatment of weldments. These address the above lifetime-limiting mechanisms in different ways depending on the type of assessment being carried out: R5, Volume 2/3—"Appendix A4. Creep–Fatigue Crack Initiation in Weldments"; R5, Volume 4/5—"Appendix A4. Creep and Creep–Fatigue Crack Growth in Weldments"; R5, Volume 6—"Assessment of Dissimilar Metal Welds"; R5, Volume 7—"Creep Rupture and Crack Growth in Similar Weldments in Ferritic Pipework".

1. Previous R5 Volume 2/3 Approach

Weldments were previously analysed in R5, Volume 2/3 (*3*) as though they were made of homogeneous parent material and the difference in behaviour of the weldment compared to the homogeneous parent material was taken into account by using a fatigue strength reduction factor (FSRF). These FSRFs were derived from fatigue tests on actual weldments by comparing the observed weldment endurance with the parent material fatigue endurance curve. The FSRF took into account the material fatigue endurance reduction,

TABLE I

FSRFs used in the previous R5 Volume 2/3 procedure to give a best estimate of fatigue endurance for austenitic steel weldments.

R5 weldment type	FSRFs	
	Dressed	Undressed (as-welded)
1*	1.5	1.5
2†	1.5	2.5
3‡	N/A	3.2

* "Type 1" = Full penetration butt weldment transverse to the main loading direction. † "Type 2" = Full penetration T-butt or fillet weldment transverse to the main loading direction. ‡ "Type 3" = Partial penetration T-butt or fillet weldment transverse to the main loading direction. N/A = Not applicable; all partial penetration welds are treated as undressed.

due to inclusions and porosity introduced in the welding process, and enhancements in strain due to material mismatch and local geometry effects. Two separate assessment routes were provided for weldments in the undressed (as-welded) and dressed conditions and FSRFs were used to enhance the strain range in order to reflect the reduced endurance of the weldment. FSRFs for the different types of dressed and undressed weldments were provided in R5, Volume 2/3, "Appendix A4" (*3*) for both austenitic and ferritic weldments. Table I shows the values that were used in conjunction with the mean parent material fatigue endurance curve to give a best estimate of the fatigue endurance of austenitic weldments. R5, Volume 2/3, "Appendix A4" (*3*) also provided FSRFs, again applied to the mean parent material fatigue endurance curve, that could be used to estimate lower bound weldment endurance.

In creep–fatigue situations, the creep damage was evaluated using the stress at the start of the creep dwell or hold period together with the predicted stress relaxation behaviour and a ductility exhaustion approach. For undressed (as-welded) weldments, where the detailed surface geometry is not known, the FSRF was also used as a strain concentration factor in determining the start-of-dwell stress.

However, this approach is considered to be overly pessimistic since the FSRF also includes the reduction in material fatigue endurance, which does not affect the start-of-dwell stress. This limitation in the previous approach provided the motivation for developing the current approach for assessing creep–fatigue initiation in weldments, which is described in the following section.

2. Current R5 Volume 2/3 Approach

The current Volume 2/3, "Appendix A4" approach for assessing creep–fatigue initiation in weldments (*2*) was primarily developed to remove known conservatisms in the

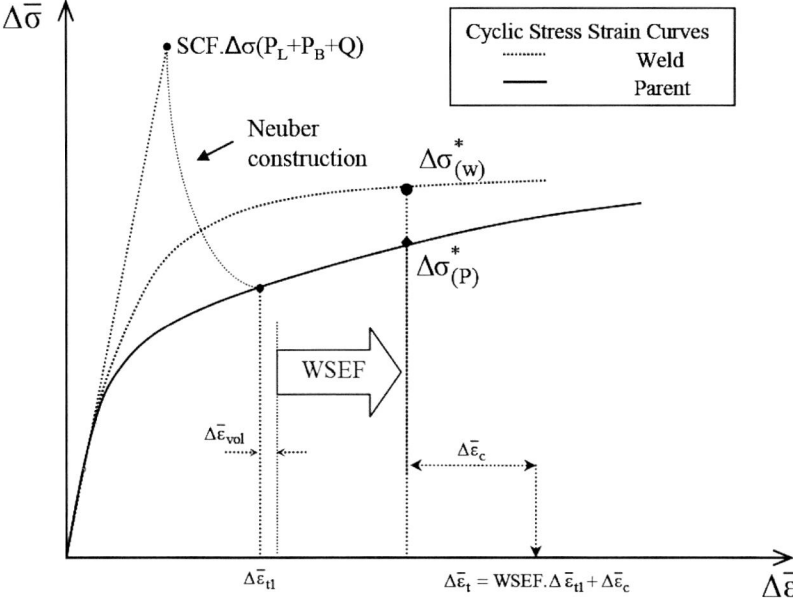

FIG. 1. — Schematic illustrating the determination of the assessment parameters in the current R5 Volume 2/3 approach.

1. For "Type 1" welds, it is not necessary to apply an SCF for weld cap angles (i.e. between the weld cap and the component surface) up to 30°. The SCF is only required for "Type 2" and "Type 3" welds with weld cap angles in excess of 39° for dressed welds and in excess of 30° for undressed welds (see footnotes to Table II).
2. $\Delta\dot{\sigma}_{(P)}$ and $\Delta\dot{\sigma}_{(W)}$ are the cyclic stress ranges calculated for assessment locations in the parent and weld metal respectively. These stress ranges are then used in conjunction with parent values of $K_s S_y$ to determine start-of-dwell stresses for locations in parent and weld metal, respectively.
3. The example above shows $\Delta\dot{\sigma}_{(P)} > \Delta\dot{\sigma}_{(W)}$. If $\Delta\dot{\sigma}_{(P)} < \Delta\dot{\sigma}_{(W)}$ then the start-of-dwell stress, σ_0, is taken as that for the parent material irrespective of whether the assessment location is in the weld or parent material.

previous approach that resulted from using the FSRF to calculate the start-of-dwell stress, and hence, creep damage.

The current approach separates the previous FSRF into the following two components: (i) the weld strain enhancement factor (WSEF), which accounts for strain enhancement due to the local geometry effects of the weldment (if applicable) and the material mismatch between weldment zones; and (ii) the weld endurance reduction (WER), which accounts for the fatigue endurance reduction due to the presence of small imperfections (e.g. inclusions, porosity, etc.) in the weldment constituent materials.

In addition, the current approach has been simplified by adopting a single route for both dressed and undressed (or as-welded) weldments. In common with the previous approach, the current approach does not account for any differences in behaviour of thick (multipass) and thin (1 or 2 bead) weldments.

The main steps in the current procedure are illustrated in Fig. 1. The first step is to perform a shakedown analysis (e.g. using finite element methods) and then to derive the linearised elastic stress range based on the primary and secondary stress ranges, $\Delta\bar{\sigma}(P_L+P_B+Q)$. This is used, in conjunction with the Neuber construction, to derive the corresponding elastic-plastic strain range using the parent material cyclic stress–strain curve. The elastic-plastic strain range is enhanced by an additional strain range, $\Delta\bar{\varepsilon}_{vol}$, to account for the difference in effective Poisson's ratio between elastic and inelastic conditions. A WSEF is then applied to the elastic-plastic strain range, which is used in conjunction with the parent material fatigue endurance curve reduced by the WER, to derive the number of cycles to crack initiation in the fatigue damage calculation. The combined effects of the WSEF on the strain range and the WER on the fatigue endurance in the current approach result in similar values of fatigue damage to those obtained using the FSRF in the previous approach.

Some additional features of the current approach are described in more detail below.

(1.) If required, a stress concentration factor, SCF (see footnotes to Fig. 1), is applied to the linearised elastic primary plus secondary stress range, $\Delta\bar{\sigma}(P_L+P_B+Q)$. Use of a SCF

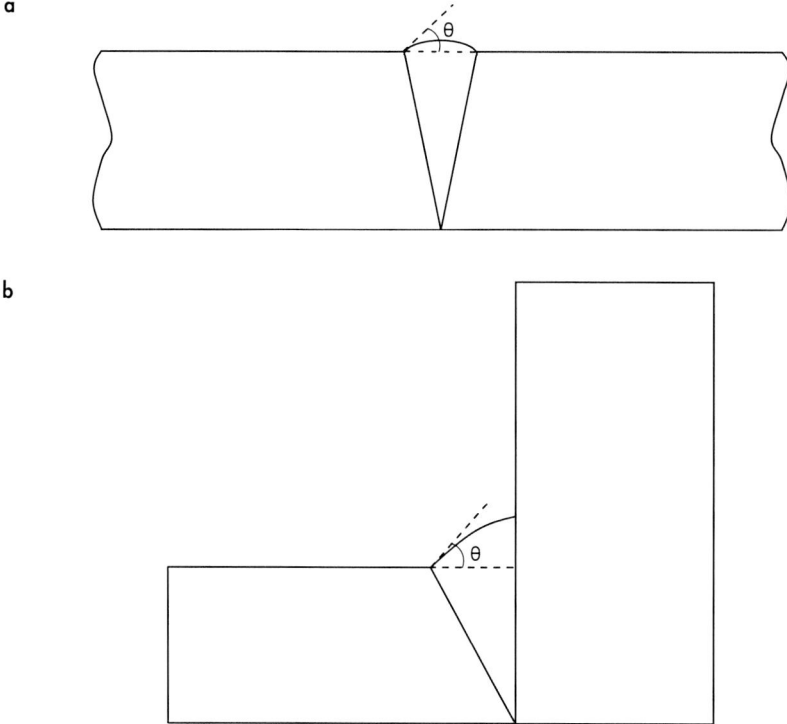

FIG. 2. — Schematic illustrating how the weld cap angle is defined. *a.* "Type 1" weld. *b.* "Type 2" and "Type 3" weld.

is only required in the procedure for weldments with more severe stress concentrations than those used to derive the WSEFs and those considered in the validation calculations. For 'Type 1' welds, it is not necessary to apply an additional SCF for weld cap angles up to 30° (see Fig. 2 a). For 'Type 2' and 'Type 3' welds, SCFs of 1.14 are already covered in the WSEF derivation and validation calculations; this means that for weld cap angles up to 30° (see Fig. 2 b) for undressed welds and up to 39° for dressed welds, no additional SCF is required. However, for 'Type 2' and 'Type 3' welds with larger weld cap angles, it is necessary to include a SCF (see Table II, footnote number 4).

(2.) An inelastic strain range $\Delta \bar{\varepsilon}_{tl}$, for the enhanced linearised elastic primary plus secondary stress range, $SCF.\Delta \bar{\sigma}(P_L+P_B+Q)$ is determined by using a Neuber construction and then adding the volumetric correction, $\Delta \bar{\varepsilon}_{vol}$. Both the primary and secondary stresses are included in evaluation of a secant modulus and effective Poisson's ratio which determine the volumetric correction.

(3.) For austenitic weldments, the inelastic strain range is multiplied by the WSEF from Table II. For ferritic weldments, different, interim WSEF values are provided in "Appendix A4" of R5 Volume 2/3 (2).

(4.) Calculation of the start-of-dwell stress, σ_0, includes the WSEF (see Fig. 1). When the dwell is at the cycle peak, the enhanced fatigue strain range, $WSEF.\Delta \bar{\varepsilon}_{tl}$ (calculated in the above step, which excludes the creep term, $\Delta \bar{\varepsilon}_c$) is used to determine the appropriate starting stress. The weldment stress range, $\Delta \overset{*}{\sigma}_{(P)}$, follows from the parent cyclic stress–strain data at the enhanced fatigue strain range (see Fig. 1). The start-of-dwell stress, σ_0, for the parent material is then obtained using parent material yield properties and the approach of "Section A7.5" in R5 Volume 2/3 (2). For an assessment point within the weld metal, if the cyclic strength of the parent material is higher than that of the weld metal, σ_0 is taken to be the start-of-dwell stress for the parent material. However, if the cyclic strength of the weld metal is higher, the parent stress range and hence start-of-dwell stress, σ_0, should be multiplied by the ratio of weld to parent material cyclic strengths at $\Delta \bar{\varepsilon}_t$ (but excluding $\Delta \bar{\varepsilon}_c$ as illustrated in Fig. 1), if sufficient weld metal cyclic stress–strain data exist. In the absence of suitable weld metal cyclic stress–strain curves, it is conservative to use the ratio of the 0.2% proof stresses of weld metal and parent from cyclic stress–strain properties to approximately account for the effect of the increased strength of the weld.

(5.) The increase in strain during the dwell due to creep, $\Delta \bar{\varepsilon}_c$, is then determined using the start-of-dwell stress, σ_0, which takes account of the enhanced strain range due to the WSEF. These calculations should use the start-of-dwell stress, σ_0, and creep properties for parent material.

(6.) The strain range, for the assessment of fatigue damage, is then obtained by adding the creep strain, $\Delta \bar{\varepsilon}_c$, determined from the previous step to the strain range without creep, $WSEF.\Delta \bar{\varepsilon}_{tl}$: $\Delta \bar{\varepsilon}_t = WSEF.\Delta \bar{\varepsilon}_{tl} + \Delta \bar{\varepsilon}_c$.

(7.) The fatigue damage is calculated using the parent material fatigue endurance curve reduced by the WER, or the weld metal fatigue endurance curve, whichever is lower.

The WER takes account of the reduction in fatigue endurance due to the presence of small imperfections (inclusions, porosity) in the weldment constituent materials and is evaluated by removing the number of cycles to nucleate a crack of depth 0.02 mm from the

TABLE II

WSEFs used in the current R5 Volume 2/3 procedure to give a best estimate of fatigue endurance for austenitic steel weldments.

R5 weldment type	WSEF
1	1.16
2	1.23
3	1.66

1. Best estimate fatigue assessments are based on the lower of the parent material fatigue endurance curve reduced by the WER and the weld metal fatigue endurance curve.
2. Bounding fatigue assessments are based on either the lower bound parent material fatigue endurance curve reduced by the WER or the lower bound weld metal fatigue endurance curve, whichever is lower.
3. For weldments with an undressed weld toe present and nominal plate thicknesses, t, greater than 25 mm and up to 150 mm, the WSEFs in Table II should be multiplied by $(t/25)^{0.25}$; noting that the factor of $(t/25)^{0.25}$ relates to fatigue cracking from the toe of a weld and, therefore, is not required for weldments without this type of detail; also, for a weldment with a weld toe present, the thickness adjustment is only necessary when assessing the weld toe location itself.
4. For "Type 1" welds, it is not necessary to apply an SCF for weld cap angles (i.e. between the weld cap and the component surface) up to 30°. For "Type 2" and "Type 3" welds, the linearised stress should be factored up by a SCF for weld cap angles (i.e. between the weld cap and the component surface, as shown in Fig. 2 b), $\theta > \psi$, where:
SCF = $\sqrt{\theta/\psi}$,
$\psi = 39°$ for dressed welds, and
$\psi = 30°$ for undressed welds

fatigue endurance. It should be noted that in R5 Volume 2/3, the fatigue endurance is divided into a nucleation phase and a crack growth phase, where the latter can be adjusted to allow meaningful assessments to be performed for thin-section structures. The WSEF is derived from the same weldment data base as the FSRF but is derived relative to the reduced endurance curve (i.e. the baseline fatigue curve factored by the WER, as shown by the intermediate dashed fatigue curve in Fig. 3); consequently, for fatigue life predictions, the combination of the WSEF and WER in the current procedure broadly corresponds to the FSRF used in the previous approach. The WSEFs in Table II have been derived such that the parent mean fatigue curve, factored by the WSEF and WER, provides a mean best fit to the weldment fatigue data assuming a log normal distribution of fatigue life.

In the previous route for undressed weldments with a creep dwell at the peak strain in the cycle, the start-of-dwell stress, σ_0, was derived by entering the cyclic stress–strain curve using the same strain range as is used in the fatigue life prediction (i.e. after the FSRF had been applied). However, in the current route, the FSRF is replaced by the lower WSEF in the derivation of the start-of-dwell stress, resulting in a lower start-of-dwell stress and hence less creep damage than was obtained using the previous route.

III. TAGSI Subgroup NT-26 Observations

The TAGSI Subgroup NT-26, was asked by EDF Energy to review the current R5 Volume 2/3 procedure for assessing creep–fatigue crack initiation in weldments. At the time of writing the subgroup report is still in draft and formal recommendations have not yet been made. However some of the more significant observations on the current methodology are discussed herein.

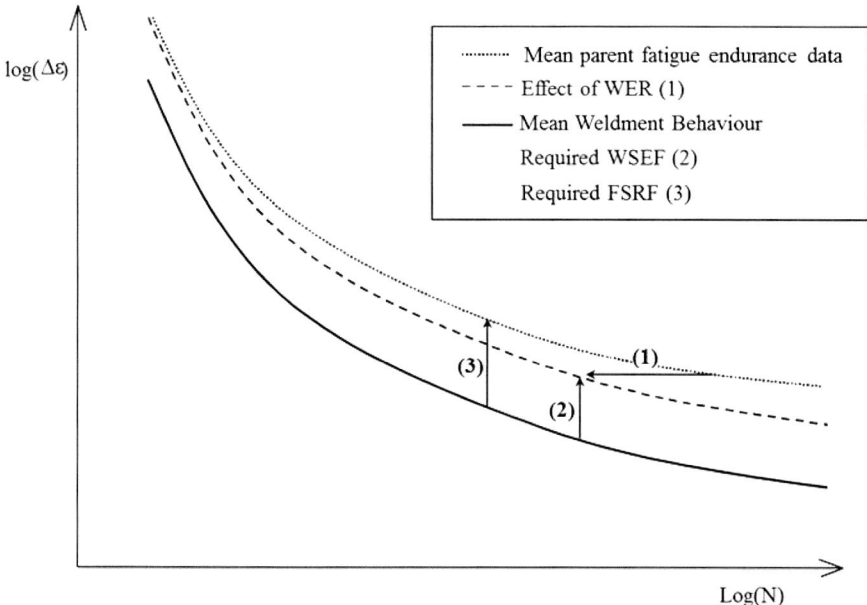

Fig. 3. — Schematic showing the effect of the original FRSF and the current WSEF and WER on parent fatigue endurance for different strain ranges.

1. Use of Fatigue Endurance Data to Determine Creep Response

As discussed in Section II.2, the WSEF is based upon fatigue endurance data of features (fully representative weldments tested at high temperature). However, when applied to plant assessments it is often the effect of the WSEF on the creep start-of-dwell stress that has the greatest impact. Whilst some creep–fatigue tests were also completed on features tests, there are insufficient tests to allow them to be used directly in developing a generic assessment methodology and therefore, they are considered only as validation of the full assessment methodology. Hence, empirical data from fatigue tests are being used to predict creep–fatigue crack initiation that may be creep dominated.

It was observed that other assessment procedures, including RCC-MRx (5), ASME Section III, Division 5 (6) and even R5 Volume 6 (2) all make use of cross-weld monotonic creep rupture strength. The cross-weld rupture strength is then used in a time fraction creep damage model to predict creep damage during a dwell. Unlike these other procedures, R5 Volume 2/3 uses ductility exhaustion creep damage models. The benefits of this approach have been the subject of another TAGSI review (7), which concluded that ductility exhaustion methods for creep–fatigue crack initiation do provide improved estimates of creep damage, especially for austenitic stainless steels.

Cross-weld uniaxial creep specimens could fail in multiple locations, parent, weld, HAZ, or fusion boundary. Due to plastic and creep mismatch between these regions, it would be difficult to determine the local creep ductility of the weldment zones which would be relevant to a creep–fatigue assessment. This makes cross-weld tests hard to utilise if a ductility exhaustion model remains the preferred route for predicting creep damage. Furthermore the cross-weld creep tests which have been completed often fail in either the parent or weld metal, away from the location of cracking in plant situations, which is often in the HAZ near the weld toe. However, the recent advent of creep and creep–fatigue testing with digital image correlation (DIC) raises the potential for monitoring local creep strains at the position of failure in cross-weld creep tests or the position of crack initiation in components.

It is therefore still believed that the best source of determining the stress–strain enhancement at the weld toe, which is relevant to the creep response, remains the fatigue data for weldment features tests. However, it is noted that whilst the fatigue data can be used to indicate the strain enhancement from geometry and plastic mismatch, it does not quantify any strain enhancement due to creep mismatch.

2. Separation of WSEF and WER

The partitioning of the effects of the WSEF and WER as described in Section II.2 and demonstrated in Fig. 3 has thus far been based on the definition of the WER relating to the nucleation of a crack of depth 0.02 mm. This value of 0.02 mm was chosen as it was a small defect size for which experimental work had been completed to separate nucleation from short crack growth in homogeneous specimens (8). In other weldment procedures such as BS 7608 (9) weldment defects are often considered to be significantly bigger, typically 0.1 mm.

Therefore it is possible that the effects of pre-existing defects are significantly underestimated, i.e. the WER is underestimated. This would result in an overestimate of the WSEF. For the assessment methodology this is a conservative assumption, as the combination of WSEF and WER should always give similar values of fatigue damage, and overestimating the WSEF is conservative for creep damage. However, further work could consider whether the current approach is overly conservative.

3. Effects of Temperature and Loading Type on WSEF Values

As presented in Table I, for austenitic stainless steels, there are currently only three WSEF values provided for the three weldment types in R5 (see the footnote to Table I). Single WSEF values are given for each weldment type that are independent of temperature and loading type. As discussed in Section II.2 the WSEF accounts for the weldment features which give rise to strain enhancement, i.e. the geometry and material mismatch effects.

It would be expected that this strain enhancement could be a function of temperature as this may change the material property mismatch. However, the available data for weldments are limited such that the effect of temperature on WSEF is not clear and therefore not considered. Similarly, whether loading is bending or membrane dominated and whether loading is displacement or load controlled would also be expected to influence the effects of geometry and mismatch on strain enhancement. However, as the available data for weldments are limited and are dominated by data from displacement controlled pure bending tests, it is hard to distinguish the effects of loading type on WSEF values.

Whilst not fully representative, inelastic finite element analysis of weldments with representative geometric features and material mismatch could be explored under different loading conditions to gain an appreciation of whether significant differences in strain enhancement due to variations in loading types and magnitude would be expected. Preliminary results do show that loading type (i.e. membrane versus bending) does affect the strain enhancement.

4. Response of Thin Versus Thick-Section Weldments

Part of the question asked of the TAGSI subgroup was to consider the future development of the procedure to consider how thin- and thick-section weldments could be considered differently. The definition and differences between 'thin' and 'thick' weldments for austenitic stainless steels are considered as follows.

(1.) *Thin-section weldments* typically contain one or two weld beads, with minimal hardening of the weld metal or HAZ due to thermomechanical cycling from the welding process. As a result the weldment zones are quite homogeneous in tensile properties, with all regions cyclically hardening during service loading. The weld cap in thinner sections tends to be a greater proportion of the thickness than in thicker sections. In addition the weld cap or root geometry can be highly variable and potentially very severe. With a small weld volume it may be expected that potential defects are smaller.

(2.) *Thick-section weldments* are typically multipass with a significant number of weld beads, resulting in cyclic hardening of the weld metal and HAZ due to the thermomechanical cycling from the welding process. This results in high levels of tensile mismatch in the different regions, which evolve as the material cycles in service (e.g. austenitic parent material cyclically hardens while multipass austenitic weld material cyclically softens). Weld caps are typically better controlled due to multiple passes, with the weld cap height being a much smaller proportion of the weld thickness than in thin-section welds. With a large weld volume it may be expected that potential defects are larger.

There are clearly differences in the behaviour of thick- and thin-section weldments that suggest that the WSEF and WER would be expected to differ. In the case of the WSEF, there are competing effects due to the potential for more onerous stress concentrating features but less significant mismatch in thin-section weldments comparted with thick-section weldments. The relative magnitude of these contributions could result in WSEFs for thin-section weldments being more or less severe than for thick-section weldments. Although the WER is likely to be more onerous for thick-section weldments due to the potential for larger imperfections or initial defects than in thin-section weldments, this effect may be offset by the larger initiation defect size likely to be used when assessing thicker weldments.

Although, in principle, there is no reason why different WSEF values could not be derived independently for thin- and thick-section weldments, limitations in the availability of suitable weldment fatigue data may preclude this approach being adopted in practice.

IV. Conclusions

Weldments are features which are prone to high levels of creep and creep–fatigue damage for a number of reasons, including the introduction of small defects, deleterious geometry, degradation of material properties (e.g. HAZ), and introduction of material mismatch. For creep–fatigue crack initiation assessments of similar metal welds, a dedicated assessment route is provided by R5 Volume 2/3, "Appendix A4". The current and previous assessment routes have been described in this paper.

Unlike the previous FSRF, which accounted for all of the deleterious aspects of weldments, the current approach uses two factors to account for specific physical attributes of the weldment. The WER accounts for the effects of small defects in weldments on fatigue endurance. The WSEF accounts for strain enhancement due to weldment geometry and material mismatch. Therefore within the R5 assessment methodology, both factors influence the fatigue life, but only the WSEF influences the creep response. As the WSEF is less than the FSRF, the current methodology results in lower predicted creep damage than the previous methodology.

TAGSI Subgroup NT-26 has been tasked with reviewing the R5 Volume 2/3, "Appendix A4" assessment route. Whilst the subgroup review is ongoing, the following topics have been highlighted as discussion points:

- the use of fatigue data to determine the WSEF, which is used to determine the enhancement of creep within a weldment
- the arbitrary separation of the WSEF and WER
- the independence of the WSEF with temperature and loading type
- expected differences in WSEF and WER between thin- and thick-section weldments

Whilst potential development opportunities for the assessment route have been highlighted within the paper, the full TAGSI Subgroup NT-26 report will provide a more complete set of recommendations on the future development of the procedure.

References

1. D. W. Dean and M. J. Chevalier, *in* "Materials and Methodology Challenges for Future Nuclear Power Plant" (R. A. Ainsworth and P. E. J. Flewitt, eds), pp. 43–56. EMAS Publishing, Warrington, UK, 2017.
2. EDF, "R5—Assessment Procedure for the High Temperature Response of Structures", Issue 3, Revision 2. EDF Energy Nuclear Generation, Gloucester, UK, 2014.
3. Nuclear Electric, "R5—Assessment Procedure for the High Temperature Response of Structures", Issue 3. British Energy Generation, Gloucester, UK, 2003.
4. CEGB, "R5—Assessment Procedure for the High Temperature Response of Structures", Issue 1. Central Electricity Generating Board, London, UK, 1990.

5. AFCEN, "RCC-MRx—Design and Construction Rules for Mechanical Components of High Temperature, Experimental and Fusion Nuclear Installations", AFCEN No. 2015-171. AFCEN, Paris, France, 2015.
6. ASME, "ASME Boiler and Pressure Vessel Code, Division 1 — Subsection NH, Class 1 Components in Elevated Temperature Service." American Society of Mechanical Engineers, New York, NY, USA, 2015.
7. G. A. Webster, A. C. F. Cocks, J. F. Knott, et al., "TAGSI Response to British Energy Question on Improved Methods for the Calculation of Creep Damage", TAGSI Report: TAGSI/P907-192. Technical Advisory Group on the Structural Integrity, Warrington, UK, 2008.
8. A. Pineau, "High Temperature Fatigue Behaviour of Engineering Materials in Relation to Microstructure, in Fatigue at High Temperature" (R. P. Skelton, ed.), pp. 305–64. Elsevier Applied Science, London, UK, 1983.
9. BSI, "BS 7608:2014—Code of Practice for Fatigue Design and Assessment of Steel Structures." British Standards Institution, London, UK, 2014.

Strengths and Weaknesses of the Proof Pressure Test Argument in RPV Structural Integrity Assessments: a TAGSI View

S. J. Garwood

Abstract. — In March 2003 TAGSI Subgroup NT11 reported on the value of proof pressure test arguments. The potential benefits of proof testing to the structural integrity of pressurised components in general are reviewed in this unpublished report. The main focus of the subgroup review was the potential benefit of the proof test in support of the individual legs of an Incredibility of Failure safety case and mitigation against in-service degradation. The NT11 report also describes the limitations of proof test arguments and reviews the approaches adopted by R6 "Revision 4" and BS7910.

This paper outlines the role of the proof test and the potential benefits of warm pre-stressing. It provides a summary of the general findings described in the NT11 report to assess the value of the proof pressure test argument in reactor pressure vessel (RPV) structural integrity assessments.

Executive Summary

TAGSI addressed two questions:

(1.) Assess the value of the proof pressure test argument in reactor pressure vessel structural integrity assessments.

(2.) Review the strengths and weaknesses of the procedures in R6 Section III.10 in relation to reactor pressure vessel structural integrity assessments undertaken after a period of service.

The above issues were considered by an ad hoc subgroup under the chairmanship of Dr S. J. Garwood. The subgroup also reviewed proof pressure test arguments relevant to a wide range of vessel types and applications. Following the work of the subgroup and deliberations of the main committee, TAGSI reached detailed conclusions in respect of these questions.

This paper outlines the general conclusions from the sub group findings.

◆

TAGSI recognises that survival of the proof pressure test, when supported by fracture mechanics analysis, can provide some assurance regarding the integrity of a pressure vessel and its fitness for purpose for future operation. Whilst relevant to a wide range of vessel types the benefits derived from survival of a proof test will vary from case to case dependent on the circumstances and the quality of the information. The proof

---

Department of Mechanical Engineering, Imperial College London, UK.

pressure test is of greatest value when the loading is higher, in the same direction and of the same type as subsequent service loadings. However, potential differences between the stressing condition at the time of the proof pressure test and that in service could undermine the potential benefits derived from PPT, if high thermal stresses coincide with temperatures at which cleavage fracture might be of concern.

The main potential benefits of the proof pressure test are:
- to provide some confirmation that the intent of the quality assurance programme employed in the construction of the pressure vessel has been realised
- to provide stress relief in welded regions and to promote the development of compressive plastic zones at local stress raisers including any pre-existing defects
- to discount the presence of certain defect size and fracture toughness combinations—effective when fracture toughness is relatively low at the proof pressure test temperature

In addition, the use of the repeat PPT may be used to mitigate against in-service degradation and (or) detrimental WPS effects.

The value of the proof pressure test in nuclear reactor pressure vessel (RPV) structural integrity assessments is limited by the following factors.
 (1.) There is potentially a wide range of defect sizes and shapes present in a given vessel at the time of the proof pressure test, and this may be difficult to assess.
 (2.) The necessary best estimate, lower and upper bound material properties that cover the full range of temperatures, times and fracture modes may not be easily defined.
 (3.) There may be an erosion of benefits due to (*i*) damage to crack-tip material or crack growth during the proof pressure test, and (*ii*) material property degradation and the occurrence of subcritical crack growth during service.
 (4.) The direction, magnitude and type of proof pressure test loading compared with subsequent operational loads, e.g. benefits of the proof pressure test do not test against cases of thermal shock loading.

The proof test should not be regarded as a stand-alone technique for assuring the future safe operation of the component. The additional confidence that it provides diminishes with improvements in the extent and rigour of the design and manufacturing quality control, quality and extent of the pre-service and in-service inspection, and efficacy of post-weld heat treatment. Survival of a proof test by an RPV can provide support to several legs of an incredibility of failure safety cases.

Probabilistic fracture mechanics analyses provide a valuable adjunct to corresponding deterministic analyses in quantifying proof pressure test benefits.

I. Introduction

Advice on the use of the proof pressure test argument for structural integrity assessments may be found in Chapter III, Section III.10 of "Revision 4" of the R6 defect assessment procedure (*1*). In particular, advice is given on the estimation of the parameter a_{proof}, which is the maximum size of defect* "which may remain in service and not have caused failure at the time of the proof test". Section III.10 calls for a number of assessment steps to be undertaken. These include:
- characterise structural geometry, defect location and size
- define the pressure and temperature corresponding to the proof test conditions
- obtain materials properties relevant to the assessment
- perform an R6 assessment using the initial estimate of defect size and best estimate material properties
- vary the defect size until failure cannot be excluded at the conditions of the proof pressure test
- take the limiting size of defect in the previous step to be a_{proof}

In essence, the above procedure uses fracture mechanics to estimate the largest defect that could have survived the proof test, or prior service loading. The intent is then to demonstrate that, allowing for in-service materials degradation and subcritical crack growth, such a defect would not be of a critical size with respect to subsequent loadings.

Against the above background, TAGSI was asked the following questions.

(1.) Assess the value of the proof pressure test argument in reactor pressure vessel (RPV) structural integrity assessments.

(2.) Review the strengths and weaknesses of the procedures in R6 Section III.10 in relation to RPV structural integrity assessments undertaken following a period of service.

In providing this response to these questions, TAGSI specifically considered steel reactor pressure vessels for nuclear applications. TAGSI reviewed the effects of proof loading both at the start of service life and periodically thereafter and has referred to TAGSI's related views on warm prestressing (WPS) and the assessment of the probability of large defects. The structure of the report is as follows. Section II provides general background to proof pressure testing in a number of industries. Section III deals with the significance of the proof pressure test in relation to the structural integrity of steel reactor pressure vessels, which is the particular application area addressed in the questions. Both deterministic and probabilistic aspects of the proof pressure test (Question 1) are addressed in Section III. Section IV reviews the strengths and weaknesses of the procedures in the R6 defect assessment procedure for assessing the proof pressure test (Question 2). TAGSI's response to the questions is presented as a set of overall conclusions in Section V.

---

* Within the context of this paper, the term "defect" is understood to mean a flaw or discontinuity that has the potential to be harmful to structural integrity.

II. General Background

The use of proof loading to provide some measure of assurance regarding the future service life of a component or structure is common to many industries. It is a general principle that the proof loading is higher than the subsequent service loading, and that the proof loading is of the same direction and type as the service loading. TAGSI has considered the use of proof loading in relation to aero engines, storage tanks, pipelines and pressure vessels, the latter including vessels with or without internal cladding. However, the main focus of TAGSI's response is the proof loading of nuclear pressure-boundary components, and in particular the reactor pressure vessel (RPV). Here, the terminology 'proof pressure test' (PPT) is normally used.

1. Temperature of PPT Relative to Normal Operating Temperature

Proof pressure tests are rarely performed at the relevant operating temperature. The PPT may be categorised into two basic types: the 'cold' PPT and the 'hot' PPT* as follows, depending upon whether the proof pressure test temperature (T_{PPT}) is less than, or greater than, the temperature for normal operation (T_{OP}).

a. Cold PPT ($T_{PPT} < T_{OP}$). Pressure vessels designed for subsequent high temperature operation are generally proof pressure tested at ambient temperature unless their toughness at this temperature is unacceptably low, in which case various codes specify the required test temperature above ambient temperature (although still below the operating temperature) (*2*). A particular example of the use of the cold PPT is in the nuclear industry, where ferritic steel reactor pressure vessels are usually tested at ambient temperature and where service temperatures are significantly higher than this. Proof pressure testing of RPVs is considered in detail in Section III.

b. Hot PPT ($T_{PPT} > T_{OP}$). For vessels designed to operate at low temperatures, the proof pressure test is generally carried out above operating temperature. For reasons of practicality, this often means testing at ambient temperature. However, in specific cases, the PPT may be carried out at a higher temperature to reduce the risk of catastrophic failure during the test. Specific examples of hot PPT include the testing of pipelines, pressure vessels and storage tanks designed to operate at low temperatures, e.g. liquid natural gas (LNG) containers. There may be a set of circumstances where in-service degradation can result in an initial cold PPT providing benefits normally attributable to hot PPT by invoking warm prestressing arguments (see Section IV.4).

2. Potential Benefits

Table I highlights the potential benefits associated with survival of a PPT, where the proof and operational loads are of the same type and in the same direction.

---

* The terms "cold hydrotest" and "hot hydrotest" are generally used in the case of water filled vessels where proof loading is applied hydraulically; however the more general terms cold PPT and hot PPT are used within the present paper.

TABLE I
Potential benefits associated with survival of a proof pressure test (PPT).

Benefit	Cold PPT ($T_{PPT} < T_{OP}$)	Hot PPT ($T_{PPT} > T_{OP}$)
(1.) Provides some assurance that a gross breakdown in quality control has not occurred	Potential qualitative benefit	Potential qualitative benefit
(2.) Provides some assurance of subsequent safe operation	Potential qualitative benefit	Potential qualitative benefit
(3.) Provides mechanical relief of residual stresses	Potential qualitative and quantitative benefits	Potential qualitative and quantitative benefits
(4.) Promotes development of compressive plastic zone at the tip of a pre-existing defect	Potential qualitative benefits (but see Section IV.4)	Potential qualitative and quantitative benefits, the latter relating to increase in effective toughness due to warm prestressing (WPS)
(5.) Excludes particular combinations of defect size and fracture toughness	Potential quantitative benefit	Potential quantitative benefit

In practice, the overall benefit of a PPT arises from a combination of the factors listed in the above table. The following paragraphs provide examples of projects that have addressed the quantification of benefits arising from the PPT.

3. Quantification of Benefits

TWI has investigated the combined effects of mechanical stress relief and warm prestressing (WPS) in relation to the PPT (*3–6*). Hydraulic pressure vessel tests, load-controlled wide-plate tests as well as large and small-scale fracture toughness tests were carried out on carbon–molybdenum and low-alloy pressure vessel steel plates and welds. In all cases the proof load was applied at a temperature that was higher than that for the subsequent test load. The main conclusions from this work may be summarised as follows.

(1.) All tests demonstrated an increase in the load bearing capacity at low temperature following a proof load at some higher temperature corresponding to temperatures near to or on the upper shelf of the fracture transition curve. It should be noted that the failure load was not necessarily greater than the proof load applied at the higher temperature.

(2.) In as-welded large-scale specimens, a substantial benefit in load bearing capacity was obtained following biaxial proof loading regardless of the temperature at which the proof load was applied. This benefit was attributable to mechanical stress relief.

(3.) In all tests, the apparent cleavage toughness following WPS in tension was higher than the as-received toughness.

A detailed review of the WPS effect, its mechanistic basis and associated quantitative models has been carried out by TAGSI and the main conclusions reported by Burdekin and Lidbury (7).

Whilst there are a number of benefits associated with survival of the PPT, there are a number of factors that could compromise these benefits including:
- damage to crack-tip material or crack growth during the PPT
- subcritical crack growth following a period of operation, e.g. fatigue, stress corrosion, etc.
- degradation of material properties following a subsequent period of in-service operation
- the effect of non-pressure loads in service.

One or more of these factors could potentially lead to vessel failure following the initial PPT, after a periodic PPT or during service. Smith and Warwick (8) have reviewed non-nuclear pressure vessel failures, and their review has recently been updated by Hayes (9). This latter study looked at over 50 pressure vessel case histories and categorised the root cause of failures into one of four types: (i) design fault; (ii) system fault; (iii) in-service degradation; and (iv) fabrication fault.

The categorisation is somewhat subjective but 36% of vessel failures were categorised as (i), 25% as (ii), 8% as (iii) and 31% as (iv). If the failures are looked at in the context of the PPT, then 31% of failures occurred during a PPT conducted either pre-service or in-service. Four percent of the in-service failures occurred in the case of vessels that had not been pressure tested. It is likely that many of the remaining in-service failures were vessels that had been subjected to a PPT. However, in each instance, a detailed analysis of the benefits of the PPT had not been performed, since there were insufficient data available.

In the pipeline industry, proof loads are commonly applied close to pressure stresses corresponding to the specified maximum yield strength of the piping material. Experience has shown that consideration should be given to the hold time on attainment of the proof pressure, since time-dependent growth of pre-existing defects may occur during the PPT. Early experience in the US indicated that 50% of failures attributable to defects occurred during initial pressurisation, a further 18% during the next four hours and a further 16% within 19 hours.

Aero engine applications do not generally use proof testing for critical parts. Integrity is demonstrated using rig tests and manufacturing control procedures. Proof loading is used however for undercarriage components. Storage tanks are filled to maximum capacity prior to service following code requirements. This provides a variable overload test to welded regions dependent on their location in a vessel.

4. Summary

The PPT is used within a number of industries that involve the operation of pressure vessels. Depending upon the particular industrial application a cold PPT or a hot PPT may be used, as appropriate. A number of benefits have been identified that arise from a pressure vessel surviving the PPT. It is evident that more quantitative descriptions of benefits due to the PPT are now available. Examples have been given where benefits due to mechanical stress relief and WPS have been investigated experimentally by observing defect behaviour in pressure vessel and associated fracture mechanics tests. However, service experience suggests that there are a number of factors that may offset PPT benefits—damage to crack-tip material or crack growth that could occur during the PPT, subcritical crack growth and (or) degradation of materials properties during a subsequent period of operation, and the effect of non-pressure loads, e.g. thermally induced stresses.

III. Proof Pressure Test in Relation to Structural Integrity of Steel Nuclear Reactor Pressure Vessels

As noted in Section II, the PPT is used in a number of industries. The types of vessel tested will include old and new vessels, clad and unclad vessels. The quality of information available to formulate a PPT argument will therefore vary from case to case and, for nuclear reactor pressure vessels fabricated from ferritic steel, some of the PPT benefits may erode as a function of time due to in-service material property changes.

For nuclear components, the ASME Boiler and Pressure Vessel Code (*10*) provides procedures for both hydrostatic (water or alternative liquid) and pneumatic (non-flammable gas) proof tests. Particular guidance is provided on the following.

a. Test temperature. It is recommended that the PPT be made at a temperature that will minimise the possibility of brittle fracture.

b. Test pressure. Components should be tested at not less than 1.25 times their design pressure. Guidance is also provided regarding the maximum test pressure to avoid gross yielding.

c. Holding time. The PPT pressure should be maintained for a minimum of ten minutes.

d. Examination for leakage. Following application of the test pressure for the required time, all joints, connections, and regions of high stress should be examined for leakage. As for pneumatic tests, the examination of pumps and valves is performed whilst the component is under a reduced pressure.

In the case of nuclear reactor pressure vessels, the PPT is used to provide assurance of future safe operation at some temperature that is higher than the proof test temperature. Therefore, the cold PPT is of relevance, since $T_{PPT} < T_{OP}$. However, it is conceivable that under start-up and shutdown conditions following a period in service, the RPV could be pressurised at temperatures (T) that are lower than the ductile-brittle transition temperature (T_{NDT}), i.e. $T < T_{NDT}$ (and $T_{NDT} < T_{PPT}$ or $T_{NDT} > T_{PPT}$, dependent on the extent of the shift in transition temperature). This would be the case if a large positive shift of T_{NDT} had occurred since the time of the proof test (i.e. due to in-service neutron irradiation). Thus, it is conceivable that there could be a greater risk of brittle fracture under pressurised conditions during routine shutdown or start-up operations. However, under such extreme conditions, it may be expected that pressure (p) would be such that $p < p_{OP} < p_{PPT}$, where p_{OP} is the operating pressure and p_{PPT} is the maximum pressure attained during the PPT. To this extent, there would be a qualitative and quantifiable benefit attributable to WPS (*7*).

1. INCREDIBILITY OF FAILURE

For a nuclear RPV, the benefits of the PPT identified in Section II for cases where the proof loading is higher, in the same direction and of the same type as subsequent operational loads, fall within three separate legs of an incredibility of failure (IOF) safety case (*11*):

a. Interpolation/extrapolation of experience (good design and manufacture). Survival of a PPT provides some assurance that the intent of quality assurance programmes employed during vessel construction has been realised.

b. Functional testing. There is a general requirement for functional testing in codes and standards. This is with the aim of providing some assurance of subsequent safe operation following survival of the test performed under appropriate loading conditions at some relevant temperature.

c. Failure analysis.

(I) FRACTURE ANALYSIS, MECHANICAL STRESS RELIEF. During a PPT the potential exists for mechanical relief of residual stresses due to the interaction between primary and weld residual stresses. This mechanism involves the combined primary and secondary stress levels attaining yield level magnitude during the test. When the primary load is removed, the remaining residual stresses will have been reduced. Simple methods are provided in BS7910 (*12*) that allow quantitative benefits to be claimed for mechanical stress relief. It should be noted that the effects of stress relief will be influenced by any post-weld heat treatment (PWHT) subsequent to proof loading.

(II) FRACTURE ANALYSIS, PLASTICITY EFFECTS. More localised plasticity effects can also occur during a PPT at stress concentration features, e.g. nozzle, weld deposited cladding, crack tip, etc. During the application of pressure loading, material may yield in tension. During subsequent unloading, material is then forced into compression by the surrounding elastic material. This may induce an advantageous compressive stress resulting in an increased resistance to failure by mechanisms such as fatigue and SCC. This benefit is enhanced by hot hydrotesting. For clad vessels, the PPT has the additional benefit of reducing the tensile residual stress remaining in the clad layer after PWHT. A special case, involving plastic deformation of crack-tip material, is the phenomenon of WPS (see below). For clad vessels, the PPT has the additional benefit of reducing the tensile residual stress remaining in the clad layer after PWHT; however, there is the possibility of some loss of the beneficial effects of compressive stresses in the underlying substrate.

(III) WARM PRESTRESSING (*7*). The phenomenon occurs when a ferritic steel structure containing a defect is loaded at one temperature and is subsequently subjected to loading at a lower temperature. It may then exhibit an increased resistance to cleavage fracture compared with that expected in the absence of preloading. The main factors believed to contribute to the effect include the following. During the preload, plasticity develops in the crack-tip region. If sufficient unloading occurs, this crack-tip plastic zone is forced into compression by the surrounding elastic material. These compressive stresses offer a level of protection to the crack tip during subsequent reloading. A reduction in temperature between the initial and subsequent loadings results in the yielded regions formed at the higher temperature becoming effectively 'frozen' into the material due to the associated increase in yield stress. This further serves to impede the development of an active plastic zone and thus inhibits cleavage.

2. Light Water Reactor Study Group

The Light Water Reactor Study Group (LWRSG) considered the PPT in 1982 (*2*) and again in 1987 (*13*). The study group came to the view that the overpressure test cannot be used to quantify the absence of specific sizes of defect. Nevertheless, they were of the opinion that it does act as a safeguard against gross breakdown of quality control and continues to have merit in terms of stress relief. Moreover, defects that survive the overpressure test could benefit from the compressive residual stress field imposed on the crack tip. However, the LWRSG considered it possible that defects just less than the limiting size to cause failure during the PPT could incur 'damage' (i.e. crack extension and [or] reduction in crack growth resistance), and that subsequently this could increase the likelihood of in-service failure.

As part of the work in support of the LWRSG, Cowan and Picker (*14*) considered deterministically the concept of a limiting size of defect that had just survived a pressure test. They did this in relation to the confidence in pressure vessel integrity that may be deduced from pressure tests. The main points arising from their work are as follows.

(1.) A limiting defect size argument must be based on the toughness values appropriate to the PPT and not lower bound values used in conventional assessments.

(2.) Consideration must be given to the possibility of crack growth by ductile tearing during the PPT.

(3.) For cases where materials are of high toughness at start-of-life and quality assurance is such that the toughness used is fully representative of the whole vessel, the possible size of remnant defect will be large. Correspondingly, the benefits of PPT arguments will be minimal.

(4.) A PPT following a period of operation in service has similar benefits and limitations as one performed at start-of-life. However, in the former case, account must be taken of changes in material properties due to in-service conditions, e.g. irradiation, thermal ageing, etc.

It is important to note, however, that all of the above work was within the context of a PWR reactor pressure vessel whose design, materials of construction and fabrication techniques were specifically intended to minimise the possibility of failure by crack initiation. In particular, the likelihood of brittle failure during the PPT at ambient temperature was remote, and critical defect sizes for the initiation of stable ductile tearing were large. Validated methods of inspection were intended to minimise the likelihood of defects of such a size escaping detection and remaining in the RPV during the PPT. Furthermore, the loadings of major concern were thermal shock loadings arising from loss of cooling, i.e. loadings very different from the proof pressure. Clearly, survival of a PPT by such a vessel can only ever provide limited information; in other words the benefits of the PPT will be small.

As a final point, TAGSI notes that the LWRSG was concerned with the integrity of PWR vessels where the thermal stress conditions of the large-break loss of coolant accident (LOCA) or pressurised thermal shock conditions are such as to make it difficult to sustain arguments regarding PPT benefits. Since publication of the LWRSG reports and the associated work of Cowan and Picker (*14*), developments in both deterministic and probabilistic fracture mechanics have taken place within a wider context than PWR vessel integrity under thermal shock transient conditions. Some of these have enhanced the interpretation that may be placed on survival of a PPT in terms of material property and defect size combinations and have been applied in interpreting the results of the PPT in the case of the RPVs of older gas-cooled reactors (*15*, *16*).

The current response considers the proof pressure test from a broader perspective than that covered by the LWRSG.

3. Interpretation of PPT Benefits

Deterministic approach. — The R6 defect assessment method provides guidance regarding the deterministic assessment of PPT benefits. A detailed appraisal of the R6 method for assessing PPT benefits is given in Section IV.

TAGSI has reviewed material property requirements for various legs of an IOF safety case (*11*) and notes that either lower bound, upper bound, or best estimate properties may be specified. The use of a PPT argument requires that material properties be defined for the time and temperature of the proof test and the in-service condition for the specific defect location of interest. This could include fracture toughness properties within the transition and upper-shelf temperature ranges. It is noted that the lower toughness properties associated with older pressure vessels potentially lead to the PPT being more discriminating than in the case of newer, high toughness vessels. This is because, in the latter case, failure associated with a critical size of defect is more likely to be a progressive ductile failure than a catastrophic event. Material property requirements for an R6 assessment are addressed in Section IV.

The specific benefit of the PPT argument revealed by fracture mechanics analysis of an RPV surviving a test is to discount the possibility of certain defect size and fracture toughness combinations. This is demonstrated by the example of the PPT assessment presented in "Annex 3", the main results of which are also summarised in Section IV.

In PPT arguments it is normal to assume that the worst defects of a given though-wall depth are surface breaking and extended in length. However, there are many other defect shapes and positions that could survive a given proof load. In a given vessel at the time of the PPT and during subsequent operation, there are potentially a wide range of defect sizes and shapes present. Due consideration should be given within a PPT argument to the results of non-destructive examination (NDE) and the detectability of relevant sizes of defect.

It is noted that since the PPT only involves pressure loading, it does not allow for the effect of non-pressure loads (e.g. thermal stresses) on critical defect size. Also, defects may grow and develop in a subcritical manner during service without breaching the pressure circuit due to a range of mechanisms, e.g. fatigue, stress corrosion cracking, etc.

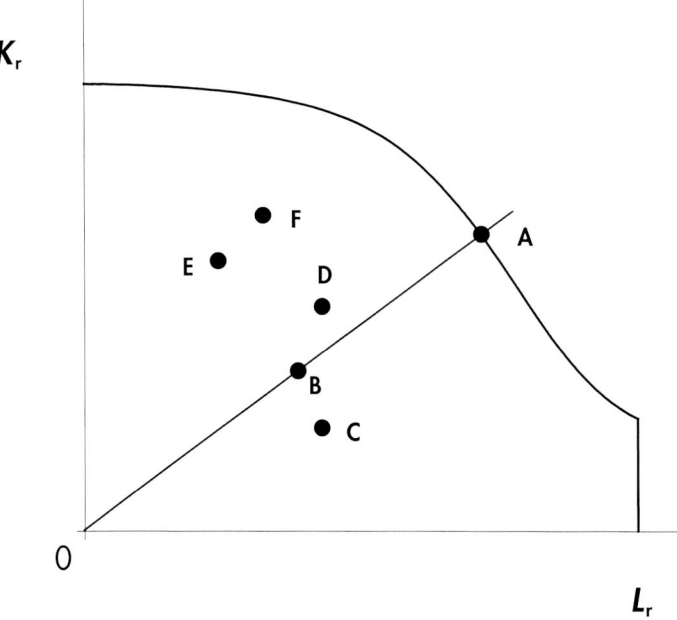

A. Assessment point at proof temperature and proof test temperature
B. Point for operating pressure and proof test temperature
C. Point at operating pressure and operating temperature
D. Inclusion of non-pressure loads
E. Change of material properties in-service
F. Inclusion of crack growth in-service

If the point **F** is within the diagram, a proof test argument can be made.

Whilst this is acknowledged in the application of a PPT argument (see Section IV), subcritical crack growth during service can reduce the PPT benefit, i.e. it can reduce the range of defect size and fracture toughness combinations that were excluded during the initial assessment. Similarly, in-service degradation of fracture toughness properties by neutron irradiation will have the effect of reducing the range of defect size and fracture toughness combinations that were excluded during the initial assessment, i.e. based on the toughness at the time of the PPT. The benefits attributable to the cold PPT ($T_{PPT} < T_{OP}$) and the way in which these benefits may be eroded is illustrated schematically with reference to the R6 failure assessment diagram in Fig. 1 as follows.

(1.) 'Point A'. This is the assessment point for the pressure and temperature of the PPT, which establishes the size of defect that may remain in the component and not have caused failure at the time of the test (a_{proof}). Materials properties (fracture toughness and yield strength) are appropriate to the temperature of the PPT.

(2.) 'Point B'. This is the assessment point for a defect of size equal to a_{proof} for a pressure equal to the normal operating pressure and with materials properties appropriate to the temperature of the PPT. Note that the pressure margin conferred by survival of the PPT is OA/OB.

(3.) 'Point C'. This is the assessment point for a defect of size equal to a_{proof} for a pressure equal to the normal operating pressure and with materials properties appropriate to normal operating temperature ($> T_{PPT}$). Note the reduction in K_r (increase in toughness) and increase in L_r (decrease in yield strength) relative to the values of K_r, L_r corresponding to 'Point B'. (Generally $K_r = K_{app}/K_{mat}$ plus a plasticity correction factor and $L_r = \sigma_{app}/\sigma_y$, where K_{app} = applied stress intensity factor, K_{mat} = fracture toughness, σ_{app} = applied stress and σ_y = yield strength)

(4.) 'Point D.' This shows the effect on 'Point C' of including non-pressure loads, which generally have the effect of increasing K_r and leaving L_r unchanged. It is evident that not including such loads may lead to a non-conservative assessment of the significance of a_{proof} and the critical defect size (a_{crit}) at the temperature and pressure of normal operation.

(5.) 'Point E'. This shows the effect on 'Point D' of in-service materials degradation leading to a decrease in fracture toughness (increase in K_r) and an increase in yield strength (decrease in L_r), e.g. as a result of neutron irradiation. It is evident that not anticipating such in-service changes in materials properties would lead to non-conservative assessments of the significance of a_{proof} and critical defect size at the temperature and pressure of normal operation.

(6.) 'Point F'. This shows the effect on 'Point E' of an increase in a_{proof} due to in-service crack growth, e.g. by fatigue (increase in K_r and increase in L_r). It is evident that not anticipating in-service crack growth would lead to non-conservative assessments of the significance of a_{proof} and critical defect size at the temperature and pressure of normal operation.

The above shows clearly how the benefits attributable to survival of a PPT may be identified, the main mechanisms for the erosion of these benefits, and how in principle these benefits may be quantified. Section IV considers these aspects in more detail within the context of the specific advice given in R6 Section III.10.

Probabilistic approach. —Whilst acknowledging the likely distributions in material properties and defect types and sizes, the above discussion has focussed on the deterministic interpretation of PPT arguments. Probabilistic treatments are addressed in the following subsection, and the specific guidance provided in R6 for both deterministic and probabilistic approaches is considered in Section IV.

The probabilistic interpretation of the PPT argument recognises that loadings, material properties, and defect depth and length will each be a distributed quantity, described by a probability distribution function (PDF) with an appropriate time-dependence. It is therefore possible, in principle, to interpret a PPT by application of probabilistic fracture mechanics. In practice, there are uncertainties in the parameters describing the input PDFs. The probabilistic interpretation of the PPT therefore requires a number of simplifying assumptions in order to make the analysis tractable. However, despite the increased complexity in analysis, a particular advantage of the probabilistic approach is that it can take account of the effects of variability in material properties and defect size in a way that deterministic approaches cannot. It is particularly effective at indicating the relative effects of specific parameters or actions even if there is uncertainty about the absolute values of the calculated probabilities of failure.

Wilson (*15*) and Connors (*17*) have described the main steps in using a probabilistic approach to quantify the effectiveness of the PPT element of a safety case. These are:

(1.) Fracture toughness and yield stress are characterised by PDFs and the failure probability based on fracture initiation for homogeneous material is calculated without taking account of the effects of any previous loadings. Defect sizes are described by PDFs for length and depth.

(2.) The analysis is repeated assuming that the vessel had survived a PPT, noting that the specific benefit of the PPT is to eliminate combinations of defect dimensions and material properties that would lead to fracture initiation during the test.

The analysis shows that the PPT reduces the failure probability of a tested vessel by a factor that depends on the form of the assumed defect size distributions. Note, this analysis does not imply that the reliability of the vessel considered will be any greater than that of the general population of vessels, since it is a mandatory requirement of all RPV design codes to carry out a PPT following fabrication of a vessel and prior to it entering service. However, overall confidence in the reliability of the vessel analysed will have improved.

The probabilistic approach is also appropriate for assessing the significance of material property correlations on PPT benefit. Correlations between material properties at the time of the PPT and the in-service time of interest, as well as spatial variations in properties at different locations in the vessel can all be considered (*15*, *17*, *18*). An extension of this is to consider the relative strength of the various legs of IOF safety cases, and how these might change with time (*11*).

4. Summary

Proof pressure test arguments are applicable to a range of nuclear pressure vessel types, including old and new vessels, clad and unclad vessels and those that undergo in-service degradation of material properties. Guidance is provided in the "ASME Boiler and Pressure Vessel Code" (*10*) regarding test temperature, pressure, holding time, and examination for leakage. The PPT is of greatest value when the loading is higher, in the same direction and of the same type as subsequent operational loads. Survival of a proof pressure test by an RPV can provide support to several legs of an incredibility of failure safety case. The main potential benefits of the proof pressure test are:

- to provide some confirmation that the intent of quality assurance programme employed in the construction of the pressure vessel has been realised
- to provide stress relief in welded regions and the development of compressive plastic zones at local stress raisers including any pre-existing defects
- to discount the presence of certain defect size and fracture toughness combinations. This is particularly effective when fracture toughness is low at the proof pressure test temperature.

TABLE II

Procedures for taking account of loading history when performing an integrity assessment of a structure [Chapter III, Section III.10 of R6 "Revision 4"].

Estimate of defect size, a_{proof}	Determination of proof test pressure
(1.) Characterise geometry, defect location, and size.	(1.) Define loads and temperatures relevant to in-service assessment, and service history between assessment time and proof test.
(2.) Define pressure and temperature corresponding to proof test condition.	(2.) Determine best estimate material properties relevant to assessment condition.
(3.) Perform R6 assessment using initial estimate of defect size and best estimate material properties.	(3.) Evaluate defect size which would just be limiting for given loads and material properties.
(4.) Vary defect size until failure cannot be excluded at proof test conditions. This defines the limiting defect size, aproof that would just survive the proof test.	(4.) Back calculate defect size to time of proof test.
(5.) Further steps to assess stability of a_{proof} in service.	(5.) Calculate the pressure required such that the defect is just limiting.
(6.) Perform sensitivity studies.	(6.) Perform sensitivity studies.

These potential benefits are affected by the following considerations. There is potentially a wide range of defect sizes and shapes that may be present in a given vessel at the time of the proof pressure test, and this may be difficult to assess. The necessary best estimate, lower and upper bound material properties that cover the full range of temperatures, times and fracture modes may not be easily defined for deterministic and (or) probabilistic assessments. Benefits may erode due to the presence of non-pressure stresses under normal operating conditions, and as a function of time due to a reduction in fracture toughness attributable to neutron irradiation and (or) subcritical crack growth.

The proof test should not therefore be regarded as a stand-alone technique for ensuring future safe operation of a component. For modern RPVs, the contribution to the overall confidence in pressure vessel integrity may be expected to diminish with improvements in the extent and rigour of the design and manufacturing quality control, the quality and extent of pre-service and in-service inspection and the efficacy of PWHT. For older vessels, the periodic PPTs may help in sustaining confidence in safety case arguments for continued operation.

Probabilistic fracture mechanics analyses provide a valuable adjunct to corresponding deterministic analyses in quantifying proof pressure test benefits.

IV. Procedures in R6 for Assessing Proof Pressure Test

Chapter III, Section III.10 of R6 "Revision 4" (*1*), describes two procedures for taking account of loading history when performing an integrity assessment of a structure which has undergone a proof or overload test. The first establishes the size of defect which would just survive the proof test (a_{proof}) whilst the second defines the maximum proof pressure for a given defect size to be just limiting. The main steps in each procedure are summarised in Table II.

An alternative method of presenting integrity arguments based on proof pressure test survival is essentially a development of the philosophy provided within R6. The main steps in the procedure are summarised in Table III.

The R6 procedure provides a structured approach and specific guidance for undertaking quantitative assessments of the PPT according to current best practice. It is noted that in this application R6 is being used as a predictive tool. Whilst there is experimental validation for R6 as a failure avoidance procedure, and for the underlying methodology in predictive mode, there remains some uncertainty regarding its predictive accuracy. Nevertheless, TAGSI considers that R6 can provide an effective method for assessing PPT benefits, conditional upon careful selection of input parameters. However, it is impractical to provide definitive validation of the PPT argument because of the wide variety of conditions which have to be considered.

There are three main aspects to consider in applying R6 to assess possible PPT benefits: (*i*) material properties, (*ii*) loading conditions, and (*iii*) defect size and type. These are likely to be different for old and new vessels due to differences in the quality

TABLE III
An alternative method of presenting integrity arguments based on proof pressure test survival.

Minimum fracture toughness inferred from survival of hydrotest

---

(1.) Define minimum defect size that will be detected during NDE, $a_{detectable}$

(2.) Assuming the presence of a defect of size defined in (1.), evaluate maximum stress intensity factor during proof test

(3.) Assuming the fracture toughness of the vessel to be equal to the value of stress intensity factor evaluated in (2.), derive the critical defect size during operation, a_{crit} (operation)

(4.) Derive the number of additional operational cycles necessary to grow a defect from $a_{detectable}$ to a_{crit} (operation)*

(5.) The component can be justified for operation up to the number of cycles (N) derived in (4.)

(6.) Perform sensitivity studies

---

* The assumption here is that the original residual stress distribution is not affected by the PPT.

of materials, in methods of construction and in the availability and quality of archive data. The procedure provides an approach in which changes in behaviour attributable to changes in crack driving force, and changes in materials resistance to crack growth can be distinguished.

1. MATERIAL PROPERTIES

R6 specifies that best estimate material properties should be selected in the calculation of a_{proof} and the determination of proof test pressure. The PPT argument is more effective when the material has lower fracture toughness at the proof test than in the assessment condition. R6 recognises that the proof test is often carried out at a temperature different from the service temperature. The effects of different material properties at these temperatures should be taken into account. Material properties may also change during service life due to thermal ageing, neutron embrittlement, WPS, or other mechanisms. The use of degraded properties would be applicable to the assessment of a_{proof} after a periodic PPT as well as the assessment of a_{crit} in service. It is noted that to maximise a_{proof}, upper bound tensile and fracture toughness properties should be used. Conversely, lower bound properties should be used to minimise the critical defect length in service (a_{crit}). However, the use of these combinations of material properties is likely to be unrealistic and lead to very conservative assessments hence the recommendation to use best estimate properties throughout. Safety justifications that include a PPT argument are therefore likely to reference different sets of material properties within different legs of the safety case.

For parts of the RPV exposed to neutron irradiation, there will be an increase in yield stress and a concurrent reduction in fracture toughness properties. It is possible to speculate that this increase in yield stress could promote the formation of residual plastic zones and hence lead to an apparent beneficial WPS effect. Although identified by Chell within the framework of his WPS model, this effect remains unproven for irradiated materials (*19*). It is noted that the complex combination of irradiation damage and stress history would make the quantification of the effects problematic. Despite this, TAGSI considers that the current treatment within R6 would provide conservative results. To take advantage of the potential theoretical WPS benefits would require appropriate validation using plastically deformed and irradiated specimens.

Whilst defects are most likely to be associated with welds, there is no explicit guidance in R6 Section III.10 regarding the treatment of over- and under-matched welds. Similarly, the effects of tearing and crack-tip constraint on fracture toughness are included elsewhere in R6 than Section III.10.

TAGSI is of the view that the current guidance in R6 regarding the selection of material properties for PPT arguments could usefully be made more comprehensive in relation to upper or lower bound behaviour, the variation of properties during service, the influence of tearing, weld mismatch and constraint effects. R6 could usefully be extended to cover these points.

R6 recommends that the significance of material property selection be assessed in one of two ways. First, sensitivity studies should be carried out to investigate the dependence of a particular assessment on the choice of material properties. An example of this approach is described in "Annex 3" for postulated external defects in pressure-boundary components, with the results summarised in Table A3.4. In this example, the assumption of a higher start-of-life toughness led to a larger value of a_{proof}, and therefore a reduced number of fatigue cycles to attain the critical defect size in service. An alternative way in which the benefits of a PPT can be presented is also provided in "Annex 3". The method involves evaluating the effective toughness of a vessel assuming that the maximum defect size that could not be detected by NDE is present during the PPT. Given that the vessel survives the PPT, the minimum toughness of the vessel can be inferred. By assuming this minimum toughness level, a safe period of service at operating conditions can be evaluated. Account was taken in the analysis of the direction and type of the proof and operational loading (including any thermal stresses).

Secondly, as noted in Section III, a full probabilistic assessment may be undertaken that incorporates material property and defect size distributions that could occur in combination (*18*). This approach, outlined by Connors (*17*) for Magnox RPVs, provides a method for quantifying a 'window' of allowable defect sizes and shapes. The probabilistic approach is complicated and requires significant investment—particularly in defining the tails of appropriate statistical distributions. The approach does provide information regarding the relative benefit of assuming different conditions, and provides a route for assessing materials that exhibit tearing resistance behaviour. Wilson (*15*) has performed a series of probabilistic fracture mechanics calculations to quantify 'worth' in IOF safety cases. (Here, worth of a safety case leg or element is defined as the logarithm [base 10] of the ratio of the failure probability without the safety case leg [or element] to the failure probability when it is included in the calculations.) The calculations illustrate the 'worth' of including the proof test leg in the safety case. Where the assessment conditions other than those relating to the initial defect distribution are the same, the actual worth of the PPT element in the functional testing leg argument is generally found to be higher for those cases where there is a higher probability of sampling smaller sizes of defect.

2. Loading Conditions

R6 requires that the loads appropriate to the PPT and the in-service condition are defined. It is noted that a PPT argument is most effective when the loading is in the same sense and of the same form as that occurring during operation, and when there is a large margin between the PPT and the assessment condition loads. When the above conditions do not apply, the procedures in R6 do not derive any benefit from the PPT. For example, in Magnox plant the high level of overpressure compared with operating and fault pressure increases the potential benefit from the PPT. In PWR plant the PPT argument is not as beneficial since the transients considered (e.g. pressurised thermal shock) provide significantly different forms of loading. Similarly, local discontinuities in thermal and structural loading within Magnox RPVs may limit the applicability of a PPT argument at some locations of the vessels.

Residual stresses should be considered at the PPT condition. The PPT is likely to modify the level and distribution of residual stresses through mechanical stress relief.

Guidance is provided in R6 Section II.7 regarding welding residual stress redistribution during the PPT.

3. Defect Size and Type

R6 requires that the initial defect size and location should be characterised as part of the PPT argument. It is noted that whilst, in general, simple defect shapes are chosen to ease the analysis (e.g. extended defects), there are many other defect sizes and shapes which may be present. These may not all be surface breaking. It is also helpful to compare the defects postulated in PPT arguments with the limiting sizes based on non-destructive examination (NDE).

For probabilistic assessments, defect size distributions are required and the result of a particular assessment will depend largely on the lower and upper 'tails' of such distributions. With reference to the latter, there are two main aspects to consider: (*i*) the probability of large defects occurring during manufacture; and (*ii*) the probability of detection of large defects by NDE. Regarding (*i*), information concerning the mechanism of defect formation can be useful in identifying a maximum defect size through modelling studies.

It is noted that in defining a PPT or periodic PPT, it might be necessary to specify the size of defect against which a given PPT is designed to guard. This could usefully place an upper bound on defect size.

R6 Section II.8 provides useful guidance regarding the calculation of the subcritical growth of defects in service following the PPT. Growth may occur by a range of mechanisms including fatigue, environmentally assisted cracking (EAC), creep and ductile tearing. It is noted that for 3D defects, subcritical crack growth may influence the shape of the defect, resulting in differences in defect geometry between the PPT and assessment conditions.

There is guidance regarding defect size and type given elsewhere in R6 for deterministic assessments of PPT benefits. This guidance could usefully be extended to reflect developments in defining the upper tails of defect size distributions, in noting time-dependent defect growth during the PPT and in highlighting potential defect shape changes during service.

4. Relevance of Warm Prestressing Behaviour to Proof Pressure Test Arguments

The WPS effect has been described in Section III, where it is noted that associated benefits occur at some temperature that is lower than the preload temperature. Warm prestressing benefits are thus normally attributable to the hot PPT. However, as noted in Section III, WPS benefits may also be applicable during shutdown or start-up operations in the case of a vessel that had received a cold PPT and had subsequently become severely embrittled by neutron irradiation.

Current models of WPS are strictly applicable to small-scale yielding plane strain situations, but their applicability may be broadened with certain caveats. In addition, within current WPS models, material property variations are only acknowledged as being caused by temperature changes. The extension to other cause — e.g. irradiation embrittlement (*19*) — is recognised in R6 though not yet validated.

Any apparent enhancement of fracture toughness due to the effects of WPS is most relevant in cases where the proof test is performed at temperatures above the ductile-to-brittle transition temperature and subsequent assessments correspond to cleavage or intergranular brittle fracture. However, it has already been noted that for nuclear reactor pressure vessels the temperature of the proof test is normally less than the subsequent service (assessment) temperature. It is still possible in theory for WPS benefits to accrue when the PPT temperature is less than the assessment temperature. However, this would only be the case if there had been a sufficiently large increase in the ductile-to-brittle transition temperature due to in-service degradation of material properties (i.e. sufficient to cause plant start-up/cool-down to occur at temperatures in the ductile-to-brittle transition range). For cases not involving a repeated PPT, service loadings at normal operating temperature may provide similar WPS benefits. It should be noted that, in this special case, any WPS benefit would in part be attributable to an associated in-service hardening causing an increase of material flow properties. However, in the absence of definitive experimental evidence, such a theoretical argument of WPS benefits remains unproven.

The above discussion relates to the effects of the PPT on cleavage toughness. However, there is also a possible effect of the PPT where the pre-strain ahead of a crack tip may induce damage that results in a decrease in the subsequent load corresponding to ductile crack initiation.

In summary, the application of WPS arguments within the context of the PPT is an area requiring further consideration in order to identify and quantify potential benefits properly.

5. Comparison with Other Approaches

In the UK, "Annex O" of the national defect assessment code, BS7910 (*12*), also considers the effect of load history with respect to the mechanical relaxation of residual stresses and the effects of WPS. No guidance is included regarding the quantification of possible benefits due to the survival of a PPT. "Annex O" recommends that if a benefit due to a PPT is claimed, then no benefit of WPS may be taken. The guidance in BS7910 regarding the assessment of WPS benefits is similar to that in R6. Both BS7910 and R6 illustrate how simplified (lower bound) and full (best estimate) WPS arguments can be made. With the possible exception of mechanical stress relief, BS7910 does not provide any advice additional to that in Chapter III Section III.10 of R6 "Revision 4".

6. Summary

The procedures in R6 Section III.10 provide a structured approach, and specific guidance for undertaking quantitative assessments of the proof pressure test and for undertaking sensitivity studies according to current best practice. Whilst there is experimental validation for R6 as a failure avoidance procedure and for its underlying methodology, it is difficult to provide definitive validation regarding its treatment of the proof pressure test. However, TAGSI considers that R6 can provide an effective method for assessing PPT benefits, conditional upon appropriate selection of input parameters. TAGSI considers that there is sufficient guidance in R6 concerning the definition of loads and assessment conditions. Guidance regarding the selection of material properties is not comprehensive. Advice could usefully be extended to cover the definition of lower and upper bound properties, the variation of properties during service, and the influence of tearing, weld mismatch, and constraint effects. Guidance regarding defect size and type is given elsewhere in R6. The latter could usefully be extended for the purpose of probabilistic assessments to reflect developments in defining the upper tails of defect size distributions and to note possible time-dependent defect growth during the proof pressure test as well as potential defect shape changes with subcritical crack growth during service. In addition, R6 could be extended to cover the influence of the PPT on mechanical stress relief.

V. Conclusions

TAGSI has addressed the following questions:

(1.) Assess the value of the proof pressure test argument in reactor pressure vessel structural integrity assessments.

(2.) Review the strengths and weaknesses of the procedures in Section III.10 of R6 "Revision 4" in relation to reactor pressure vessel structural integrity assessments undertaken after a period of service.

Proof pressure test arguments are relevant to a wide range of industries and vessel types, including old and new vessels, clad and unclad vessels, and those that undergo in-service degradation of material properties. Depending upon the particular industrial application a cold proof pressure test or a hot proof pressure test may be used, as appropriate. However, in the case of nuclear reactor pressure vessels, the cold proof pressure test is used to provide assurance of future safe operation at some temperature that is higher than the proof test temperature. TAGSI has reached the following conclusions in the case of a steel nuclear reactor pressure vessel surviving such a test:

QUESTION 1

The proof pressure test is of greatest value for steel reactor pressure vessels when the loading is higher, in the same direction and of the same type as subsequent service loadings. However, potential differences between the stressing condition at the time of the proof pressure test and that in service could undermine the potential benefits derived from the PPT, if high thermal stresses coincide with temperatures at which cleavage fracture might be of concern. Survival of a proof pressure test by an RPV provides support to the individual legs of an Incredibility of Failure safety case. The proof pressure test is particularly effective when fracture toughness is relatively low at the proof pressure test temperature. The main potential benefits of the proof pressure test are:

- to provide some confirmation that the intent of the quality assurance programme employed in the construction of the pressure vessel has been realised
- to provide stress relief in welded regions and to promote the development of compressive plastic zones at local stress raisers including any pre-existing defects
- to discount the presence of certain defect size and fracture toughness combinations — this is particularly effective when fracture toughness is relatively low at the proof pressure test temperature

In addition, the use of the repeat PPT may be used to mitigate against in-service degradation and (or) detrimental WPS effects.

The value of the proof pressure test is limited, however, by the following factors:

(1.) There is potentially a wide range of defect sizes and shapes that may be present in a given vessel at the time of the proof pressure test, and this may be difficult to assess.

(2.) The necessary best estimate, lower and upper bound material properties that cover the full range of temperatures, times and fracture modes may not be easily defined.

(3.) There may be an erosion of benefits due to (*i*) damage to crack-tip material or crack growth during the proof pressure test, and (*ii*) material property degradation and the occurrence of subcritical crack growth during service.

(4.) The direction, magnitude and type of proof pressure test loading compared with subsequent operational loads, e.g. benefits of the proof pressure test do not test against cases of thermal shock loading.

Probabilistic fracture mechanics analyses provide a valuable adjunct to corresponding deterministic analyses in quantifying proof pressure test benefits.

Question 2

The procedures in R6 Section III.10 provide a structured approach and specific guidance for undertaking quantitative assessments of the proof pressure test and for undertaking sensitivity studies according to current best practice. Whilst there is experimental validation for R6 as a failure avoidance procedure and for its underlying methodology, it is difficult to provide definitive validation regarding its treatment of the proof pressure test.

Guidance regarding the selection of material properties is not comprehensive. Advice could usefully be extended to cover the definition of lower and upper bound properties, the variation of properties during service, the influence of tearing, weld mismatch and constraint effects, and the influence of the proof pressure test on mechanical stress relief.

Survival of the proof pressure test, when supported by fracture mechanics analyses, can provide some assurance regarding the integrity of a pressure vessel and its fitness for purpose for future operation. There remains some uncertainty regarding the accuracy of the calculated benefits, since R6 has been validated as a failure avoidance procedure rather than as a predictive defect assessment procedure. However, TAGSI considers that R6 can provide an effective method for assessing proof pressure test benefits, conditional upon appropriate selection of input parameters.

◆

Guidance regarding defect size and type is given elsewhere in R6. This could usefully be extended for the purpose of probabilistic assessments to reflect developments in defining the upper tails of defect size distributions and to note possible time-dependent defect growth during the proof pressure test as well as potential defect shape changes with subcritical crack growth during service.

◆

Warm prestressing (WPS) benefits are normally attributable in the case of the hot proof pressure test, where the test temperature is higher than the temperature for normal vessel operation. However, WPS benefits may also be applicable during shutdown or start-up operations in the case of a vessel that had received a cold PPT and had subsequently become severely embrittled by neutron irradiation.

◆

BS7910 provides an alternative treatment for mechanical stress relief to that given in Chapter III Section III.10 of R6 "Revision 4". However, in all other respects, R6 provides more comprehensive guidance on the treatment of the proof pressure test.

Acknowledgements

The exceptional contribution of the late Professor David Smith to TAGSI and, in particular, the understanding of the effect of post-weld heat treatment and proof testing on residual stresses. This paper is based on the general findings of the report "TAGSI P(02) 174 NT11", March 2003, by S. J. Garwood, D. P. G. Lidbury, J. S. Schofield, and A. H. Sherry.

References

1. British Energy, "Assessment of the Integrity of Structures Containing Defects", R/H/R6, Revision 4. British Energy Generation, Gloucester, UK, 2001.
2. UKAEA, "An Assessment of the Integrity of PWR Pressure Vessels", Second Report of a Study Group under the Chairmanship of Dr W. Marshall, CBE, FRS, March 1982.
3. S. J. Garwood, K. Bell, and K. Smith, *in* "Pressure Vessel Fracture, Fatigue and Life Management", (S. Bhandari, P. P. Milella, and W. E. Pennell, eds), PVP-Volume 233, pp. 45–50. American Society of Mechanical Engineers, New York, NY, USA, 1992.
4. D. J. Smith and S. J. Garwood, *in* "Fracture Mechanics: Twenty-Second Symposium" (H. A. Ernst, A. Saxena, and D. L. McDowell, eds), ASTM STP 1131, Volume 1, pp. 833–49. ASTM International, West Conshohocken, PA, USA, 1992.

5. D. J. Smith and S. J. Garwood, *Int. J. Pres. Ves. Pip.* **51,** 241 (1992).
6. D. J. Smith and S. J. Garwood, *in* "Structural Integrity Assessment" (P. Stanley, ed.), pp. 99–110. Elsevier Applied Science, London, UK, 1992.
7. F. M. Burdekin and D. P. G. Lidbury, *Int. J. Pres. Ves. Pip.* **76,** 885 (1999).
8. T. A. Smith and R. G. Warwick, *Int. J. Pres. Ves. Pip.* **11,** 127 (1983).
9. B. Hayes, "Root Cause Analysis of Pressure Vessel Failures", TWI Report No. 12512/1/00. TWI, Cambridge, UK, 2000.
10. ASME, "ASME Boiler and Pressure Vessel Code, Section III—Rules for Construction of Nuclear Facility Components, Division 1—Subsection NB, Class 1 Components." American Society of Mechanical Engineers, New York, NY, 2001.
11. R. Bullough, F. M. Burdekin, O. V. J. Chapman, et al., *Int. J. Pres. Ves. Pip.* **78,** 539 (2001).
12. BSI "BS 7910:1999, Annex O. Consideration of Proof Testing and Warm Prestressing." British Standards Institution, London, UK, 2000.
13. UKAEA, "An Assessment of the Integrity of PWR Pressure Vessels", Addendum to the Second Report of the Study Group, since 1982 under the Chairmanship of Sir P. B. Hirsch, FRS, April 1987.
14. A. Cowan and C, Picker, *Int. J. Pres. Ves. Pip.* **15,** 105 (1984).
15. R. Wilson, "A Preliminary Investigation into the Use of Probabilistic Fracture Mechanics to Quantify 'Worth' in Incredibility of Failure Safety Cases", BNFL Magnox Division, Report TE/GEN/REP/0044/97. British Nuclear Fuels, Warrington, UK, 1997.
16. R. A. Ainsworth and R. Wilson, "Application of Probabilistic Fracture Mechanics to Pressure Vessels", International Conference on PSA/PRA for the Nuclear Industry, Café Royal London, November 1993.
17. D. C. Connors, "A Review of the Proof Pressure Test Argument Applied to Magnox Reactor Steel Pressure Vessel Safety Cases (ex-CEGB plant)", BNFL Magnox Division, Advice Note M/RS/GEN/EAN/0097/00. British Nuclear Fuels, Warrington, UK, 2000.
18. A. Muhammed, "Reliability analysis studies on welded structures", PhD Thesis. UMIST, Manchester, UK, 1995.
19. G. G. Chell and J. R. Haigh, *Int. J. Pres. Ves. Pip.* **23,** 121 (1986).

Stress Based NDE: Taking Infrared Thermography Inspection from the Laboratory to the Power Station

R. C. Tighe[1], J. P. Tyler[2], G. P. Howell[3], and J. M. Dulieu-Barton[3]

ABSTRACT.—The development and implementation of thermoelastic stress analysis (TSA) as a new stress based non-destructive evaluation approach is described. On-site tests consisted of the inspection of several welds along thick walled high pressure steam drains during a scheduled outage period at a coal fired power station. Inspections were found to be very efficient and robust in this environment with data collection and analysis taking a matter of minutes.

It is demonstrated that the equipment was robust in the difficult service environment, with data collection and analysis taking a matter of minutes identifying stress concentrations close to welds. The work shows that TSA can be used effectively as a tool to rapidly inspect defects, which could be used in conjunction with other inspection techniques such as ultrasound to reduce costs and inspection time.

I. Introduction

In the power industry, as with many others, the most common procedure used for the identification of defects in welds is typically phased array ultrasound (UT) (*1*) The process for weld inspection and reporting is laborious involving comparison to a reference case and manual plotting of results. As such, the time consuming nature of UT results in the inspection of only selected sites. Thermoelastic stress analysis (TSA) (*2*) is a thermographic non-destructive technique which uses an infrared (IR) detector to measure the surface temperatures of a component whilst it undergoes cyclic loading which is then used to create a map of the surface stresses.

The surface temperature change, ΔT, of the cyclically loaded component, is related to the sum of the principal stresses as follows (*2*):

$$[1] \qquad \frac{\Delta T}{T} = -K(\Delta\sigma_1 - \Delta\sigma_2) \quad ,$$

where T is the surface temperature, $\Delta\sigma_{1,2}$ are the principal stresses and K is the thermoelastic constant of the component material:

---

[1] Faculty of Science and Engineering, University of Waikato, Hamilton 3240, New Zealand.
[2] Enabling Process Technologies Ltd, Portishead, Bristol BS20 7AN, UK.
[3] University of Southampton, University Road, Highfield, Southampton SO17 1BJ, UK.

[2]
$$K = \frac{\alpha}{\rho C_P},$$

where α is the coefficient of thermal expansion, ρ is density and C_P is the specific heat at constant pressure.

As the results of TSA are presented in the form of images, there is no need to make drawings to estimate damage size as is typically the case for industrial UT inspection, rather the data may be directly assessed and stored. Typically TSA conducted in a laboratory setting relies upon a servo-hydraulic test machine to apply the cyclic load to create the thermoelastic response; the current work shows that excitation by means of a vibration type loading is sufficient to create the strain change required for TSA. A portable pneumatic-based loading system was developed that is capable of providing sufficient load to create a measurable thermoelastic response. The present paper focusses on the development and implementation of the portable TSA approach for stress based on-site non-destructive evaluation. Initial trials are presented showing the inspection of several welds along thick walled high pressure steam drains during a scheduled outage periods at UK coal fired power stations West Burton and Cottam (EDF Energy). The present paper focuses on the practical application of TSA, a more detailed account of the technique development and underlying physics can be found in Tighe et al. (3).

II. Towards Onsite Pipe Inspections

The first challenge of natural frequency excitation of the pipe systems was to identify a loading device capable of providing appropriate force and frequency excitation. In the case of inspecting pipes in the power station it is important that the shaker is as versatile as possible to be able to excite a range of pipes. Typical pipes found in the power stations varied in length between half a metre and tens of metres. The outer diameter (OD) also varied between 40–48 mm and wall thickness of the pipes, typically between 6–10 mm. All these variations in parameters result in required loading frequencies between a few Hz and a few hundred Hz. Another consideration when selecting a shaker is the weight of the shaker itself. The shaker must be attached to the pipe being inspected and so consideration must be given to the mass of the shaker both in terms of the additional load on the pipe as well as health and safety concern of having a heavy mass suspended potentially at height. A further detail that was considered is the power supply that is available. During the outage periods while inspection can take place the only power available was 110 V, 16 A, single phase.

One method of applying the required excitation is via a pneumatic shaker. Pneumatic shakers use the flow of compressed air to turn an eccentrically weighted turbine wheel thus creating a vibration excitation. The rate of air flow may be varied which then varies both the frequency and force produced. It should be noted that the frequency and force are coupled, i.e. the higher the frequency the higher the force produced by the turbine wheel. The pneumatic shaker selected was a GT36 from Vibratechniques Ltd

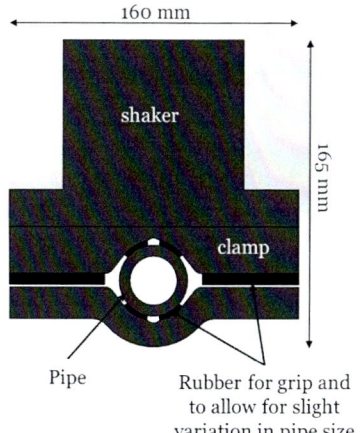

FIG. 1. — Clamping system to attach shaker system securely to pipes.

(Shoreham-by-Sea, UK). It was necessary to manufacture a clamping system to attach the shaker to the pipes. A clamp was designed that was able to accommodate a small amount of variation in the OD while being able to securely clamp the pipe to prevent the shaker spinning itself around the pipe. The clamp faces that contacted the pipe were covered in a reinforced natural rubber to allow the pipe to be securely gripped. Once the clamp is fastened tightly onto the pipe the rubber will be compressed and therefore does not cause damping between the shaker and the pipe. The clamp design and dimensions are shown in Fig. 1.

The setup for laboratory TSA application of the sample pipe is given in Fig. 2. The IR detector is a Flir SC5000, which incorporates a 256×320 indium/antimonide (InSb) photon detector array capable of recording at 383 Hz at full frame. The thermal resolution of the detector is 20 mK, which is reduced to approximately 4 mK with application of the lock-in processing used in TSA. The reference signal for the lock-in was provided using an accelerometer. The natural frequencies of pipes are checked using instrumented hammer tests where the frequency response function is analysed to find

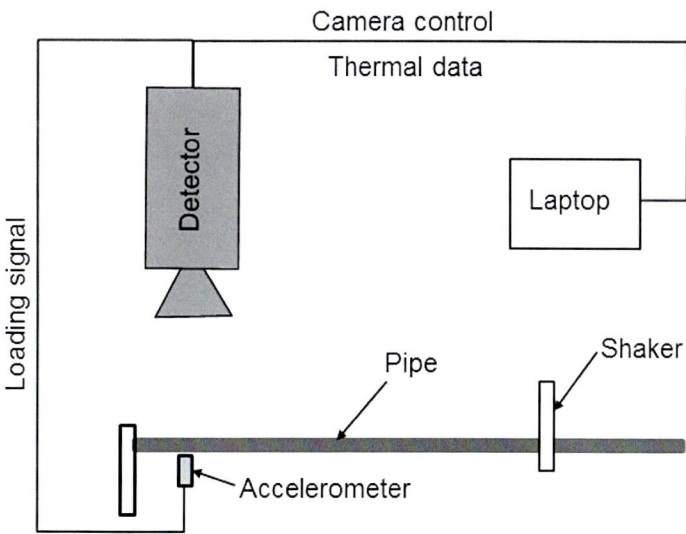

FIG. 2. — Schematic for TSA pipe inspection for laboratory.

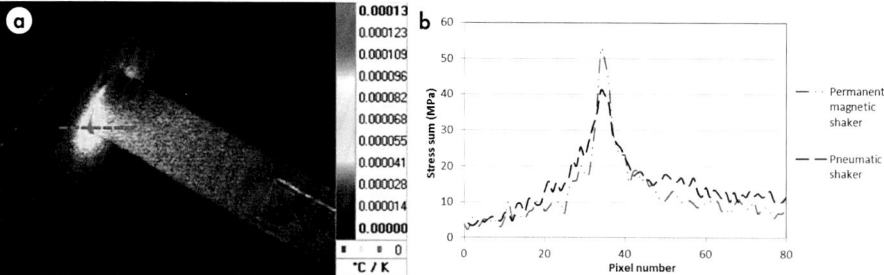

Fig. 3. — *a*. Thermoelastic response from the laboratory demonstrator. *b*. Comparison of the response using an electromagnetic shaker and the pneumatic shaker.

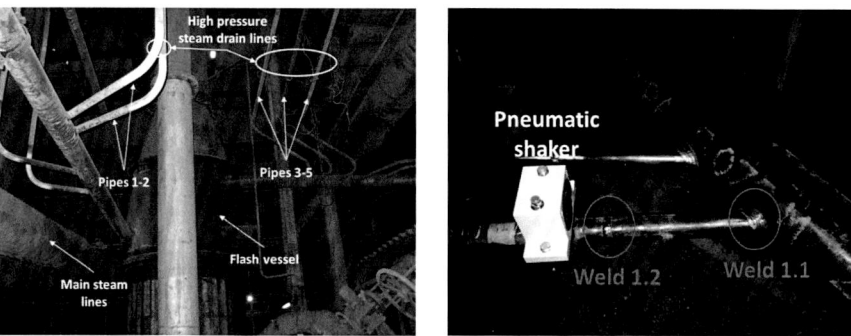

Fig. 4. — Overview of inspection area at EDF West Burton and pneumatic shaker positioned to excite pipe 1, weld 1.1, and weld 1.2.

the excitation frequency. A sample pipe of 1.5 m long was provided. Two end plates were tungsten inert gas (TIG) welded to the pipe based on the welding procedure used for steam pipes in power stations. The pipe ends by the welds were the focus of inspection for the laboratory trails. Typical results are shown in Fig. 3: Fig. 3 *a* shows the TSA image of the pipe junction and Fig. 3 *b* shows a line plot along the red line indicated in Fig. 3 *a*. The results from the pneumatic shaker are compared with the results using excitation for a standard electromagnetic shaker. There is a good agreement between the two hence it was considered that it had been demonstrated that this form of excitation could elicit the necessary thermoelastic response to make a measurement on site.

1 ON-SITE INSPECTION: WEST BURTON

Fig. 4 shows the pipes identified for on-site inspection at West Burton power station, which provided a range of weld configurations to consider. Pipes 1–2 had an outer diameter of 48 mm and joined a larger section of pipe creating a T-junction, numbered weld 1.1. Only data from pipe 1 is presented in the current paper. Along the length of pipe 1 are three further welds numbered welds 1.2 to 1.4. Pipe 1 had a 600 mm horizontal section which was welded to the larger section at one end (weld 1.1), at the other end there was a 90° bend which changed the direction of the pipe from horizontal to vertical for approximately three metres. The first two welds on pipe 1 on the horizontal section of the pipe are shown in Fig. 4.

It was necessary to prepare the surfaces of the inspection areas. Pipe 1 was newly fitted pipe so it contained little corrosion thus only a light sanding with 80-grit SiC paper was required. For older pipes it was necessary remove heavy surface corrosion prior to TSA inspection. Such welds were cleaned using a battery powered grinder with an 80-grit flapper wheel, taking around three to five minutes per weld. The surfaces were then cleaned with acetone. Such surface preparation is the same as that required for other inspection techniques, including ultrasound. A thin layer of matt black spray paint was then applied to the inspection areas to provide high and uniform emissivity for TSA inspection. Due to the time restrictions for work on site the shaker was positioned at a single site on each pipe to inspect all welds. The position for pipe 1 was located at the midpoint of horizontal section of the pipe as shown in Fig. 4. The shaker was run using a constant air pressure at 1.5 bar for all excitations, the exact excitation frequency was then determined using a self-reference lock-in process (*4*). Self-referencing means the reference signal used for the TSA processing, typically taken as the load cell signal when using a test machine, is taken from the collected IR data. The self-referencing takes the mean recorded value over an area of pixels from the surface of the component through time. The areas selected for the self-referencing are close to the weld to ensure that the frequency of the local vibration of interest is used to process the data. To extract a frequency spectra, a fast Fourier transform (FFT) is applied to the reference data; the frequency component with the largest amplitude is used as the reference frequency for the lock-in processing into TSA data. The current self-referencing approach is most suited where cyclic loading is applied however there are other self-referencing approaches in the literature, for example Tighe et al. (*3*), suited to different loading conditions.

The $\Delta T/T$ (thermoelastic response) data for welds 1.1 and 1.2 are presented in Fig. 5 *a* and Fig. 5 *b*. Stress concentrations occur at the sides of weld 1.1 with a line of zero stress

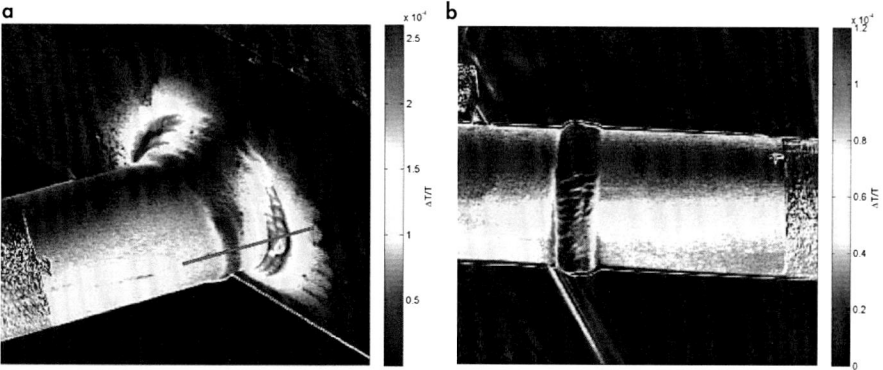

FIG. 5. — Thermoelastic response from: *a*. weld 1.1 with profile line; and *b*. weld 1.2.

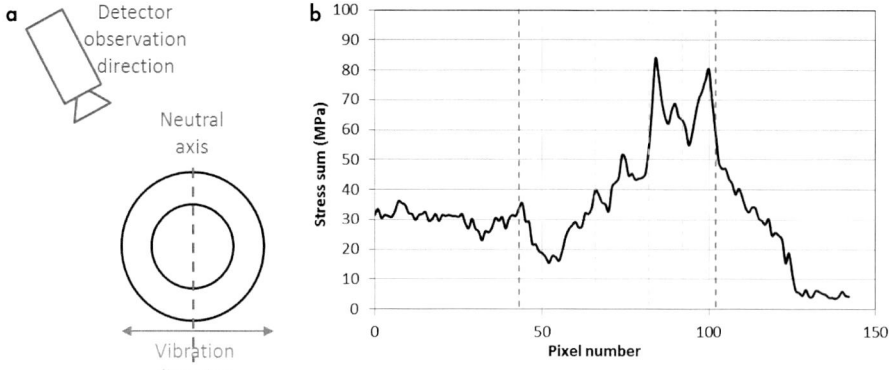

FIG. 6. — *a*. Observation orientation of the detector relative to the primary vibration direction. *b*. Stress sum data calculated along the profile line in Fig 5 *a*.

along the top of the pipe. This is attributed to the geometry of the pipe system providing more reinforcement in the vertical direction due to the vertical bend in pipe 1. The selection of the detector position is important as this governs what is observed in the TSA data. In some instances it would be necessary to collect data from multiple observation angles. The detector was positioned at around 50° relative to the vibration direction, as shown in Fig. 6 *a*. While in weld 1.1 the highest levels of stress are found at the edges of the weld material in weld 1.2, shown in Fig. 5 *b*, stress concentrations are located either side of the weld in the lower edge of the image. The effect of motion is much more apparent in the data for weld 1.2 which manifests itself as blurring at the edges outlining the pipe in the $\Delta T/T$ data. The effect of motion is also apparent at the edges of the pipe in the phase data for weld 1.2.

The maximum stress sum that the pipe is experiencing is found at weld 1.1. A profile of the $\Delta T/T$ data is taken across the weld, along the line shown in Fig. 5 *a*. Using an experimentally determined thermoelastic constant for the pipe material, it is possible to calculate the sum of the principal stresses found in the weld as plotted in Fig. 6 *b*. Either end of the extent of the weldment is marked on the graph as are the approximate locations of the edges of each weld pass with weldment edges marked in red dashed line and position of each weld pass marked as an orange dashed line. The maximum stress sum is 84 MPa. Although initially there is a reduction in stress in the first weld pass at the smaller pipe end there is a general increase in magnitude of the stress sum within the weld moving from the smaller pipe to the larger pipe. There are several peaks in the stress sum data across the width of the weld which appears to correlate with the weld passes. The work has demonstrated that it is possible to take TSA on-site and obtain useful data rapidly during outages. A key issue highlighted by the work was the effect of rigid body motion when the test specimens were under the vibrational loading and also the need to correlate better with the geometric imperfections in the pipework. The second set on on-site tests addressed both concerns.

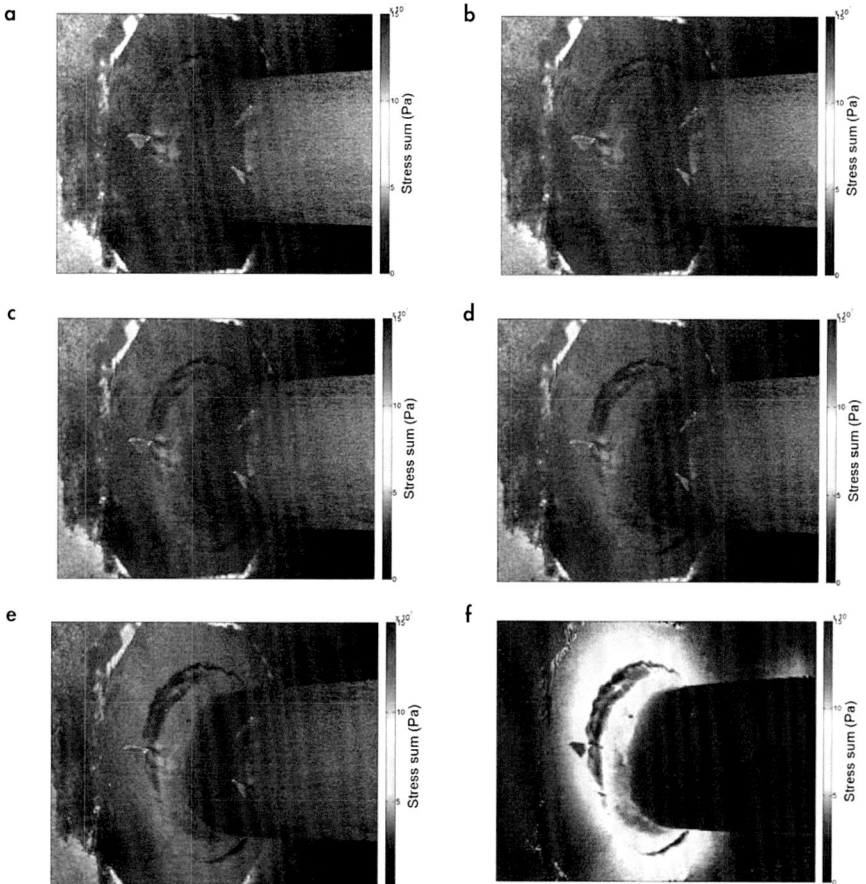

FIG. 7. — TSA stress sum data with raw IR image taken prior to loading to enable correlation between stress distributions, surface features and weld geometry. Transparency of the overlaid raw IR image is increased moving from a to f.

2 ON-SITE INSPECTION: COTTAM

Fig. 7 shows data taken from a pipe junction from the second set of trials at Cottam power station. By overlaying a raw IR image taken prior to loading (Fig. 7 a) on the TSA stress data (Fig. 7 f) and varying the transparency of the raw data, it is possible to correlate features in TSA and raw IR images to eliminate spurious results cause by surface features and correlate stress concentrations with surface geometries or location. Fig. 7 shows increasing transparency overlaid image moving from a to f. Without such correlation false positives may be caused by variations in emissivity on the surface; for example in Fig. 7 a three triangular markers are visible, used to eliminate the effect of rigid body motion. However,

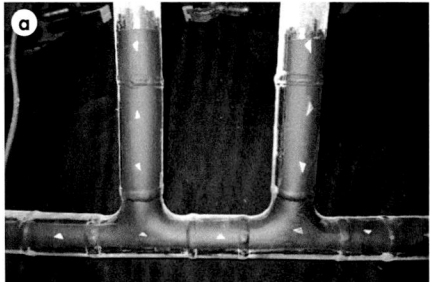 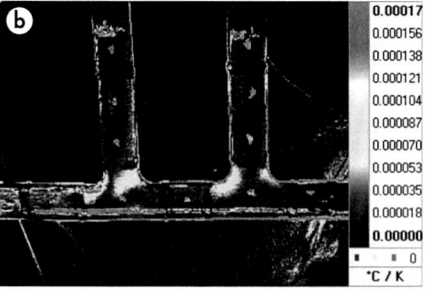

Fig. 8. — Demonstration of rapid survey capability of on double pipe junction. *a*. Painted pipework with markers for motion compensation. *b*. Raw thermoelastic response image.

these remain visible in the stress data by correlation; such false artefacts can be discounted from the data so that the stress data can be better interpreted. Application of the markers, enabled the effect most of the motion to be eliminated from the raw IR data.

An example of the speed of application of the on-site TSA defect inspection technique and its ability to rapidly assess large sections of pipework is shown in Fig. 8. Fig. 8 *a* shows a double junction in pipework prepared for the application of TSA along with the markers for motion compensation; the pipe OD was 51 mm. Fig. 8 *b* shows the thermoelastic response from the pipes in raw form as would be seen on the computer screen on-site; the stress concentrations at the junctions are immediately visible and the effect of quite significant rigid body motion has been eliminated by tracking the markers on the pipework. It is also clear that the noise at the bottom of the image is a result of the unpainted area and the motion, so further refinement of the motion compensation algorithm is necessary, which is a topic of current research.

III. Conclusions and Future Work

The present paper has shown that TSA can be conducted with a vibration loading and has been successfully applied on-site as an assessment approach. The methodology was demonstrated on-site on thick walled stream drains. On-site application using a pneumatic shaker was successful and yielded comparable results to previous laboratory trials. Further work will, firstly, review methods to reduce the effect of motion and apply these in future testing and, secondly, address the identification of known defects to characterise different types of defect known to occur in welds.

Acknowledgements

The work presented in the paper is part of an Innovate UK-sponsored collaborative project under the "Developing the Civil Nuclear Power Supply Chain Call", No. 101438, REsidual Stress and Structural Integrity Studies using Thermography (RESIST). Project partners included Enabling Process Technologies Ltd, University of Southampton, National Physical Laboratory, TWI Ltd,

EDF Energy, and AMEC Foster Wheeler (now Wood plc). In particular we thank Professor Andy Morris and Mr Stephen Lormor of EDF Energy for facilitating the on-site work and enabling the necessary safety training to provide us with access. We also thank Dr Neil Ferguson at the University of Southampton for his help and support.

References

1. A. Erhard, G. Schenk, T. Hauser, and U. Volz, *Nucl. Eng. Des.* **206**, 325 (2001).
2. J. M. Dulieu-Barton, *in* "Optical Methods for Solid Mechanics" (P. Rastogi and E. Hack, eds), pp. 345–66. Wiley-VCH, Weinheim, Germany, 2012.
3. R. C. Tighe, G. P. Howell, J. P. Tyler, S. Lormor, and J. M. Dulieu-Barton, *NDT&E Int.* **84**, 76 (2016).
4. T. Sakagami, Y. Izumi, N. Mori, and S. Kubo, *QIRT J.* **7**, 2010. (10th International Conference on Quantitative Infrared Thermography, Québec, Canada, 27th–30th July 2010.) https://doi.org/10.3166/qirt.7.73-84

Probabilistic Structural Integrity

N. A. Zentuti, J. D. Booker*, J. Hoole,
R. A. W. Bradford, and D. Knowles

ABSTRACT. — This paper highlights the wide range of applications for probabilistic analyses in structural integrity and design. Firstly, some background is given to introduce the basic principles which underpin a probabilistic approach and some aspects of its historical development. A discussion of the main benefits and attributes of probabilistic analyses follows, and a direct comparison with conventional deterministic design is made. Thereafter, five main modes of application are presented, and their individual objectives discussed. Through examining various case studies, the different utilities of a probabilistic approach are highlighted. Finally, some important considerations are discussed in terms of adopting probabilistic approaches for future implementation.

I. Introduction

Probabilistic approaches for design and structural integrity analysis have been with us for over 50 years now and have been found to facilitate a more realistic understanding of performance of components, products, and systems through the incorporation of uncertainties in input parameters. Historically, these uncertainties have been catered for using large factors of safety (sometimes termed factors of ignorance) in a deterministic approach. A deterministic approach may result in inconsistent designs (over- or under-design) owing to limitations in the information about design, manufacturing, material, and service related parameters (*1*). Conversely, in a probabilistic approach, methods are used to investigate the combination and interaction of these input parameters, having characterised distributions, to estimate the probability of failure, for optimisation and to perform sensitivity studies. Probabilistic approaches are said to be the only established methodology of propagating uncertainty through engineering models (*2*), yet they have not been taken up routinely by industry, and a deterministic culture still dominates many engineering domains to date.

A brief review of the use of factors of safety in practice is an important step in understanding why probabilistic approaches were developed. The explicit use of safety factors in calculations started circa 1850s and continued unchallenged for about 100 years (*3*). Lower limits of strength and upper limits of loading stress were typically applied in anticipation of uncertainty, but with safety factors applied on top of these judgments to account for other uncertainties associated with unknown factors, in particular, for

---

Solid Mechanics Research Group, Faculty of Engineering, University of Bristol, Bristol, UK.
* EMAIL: j.d.booker@bristol.ac.uk

modelling inaccuracies. Safety factors were also developed for different areas of application as domain specific expertise and experience guided (*4*). As engineers learned more about the nature of variability in engineering parameters in general, but were unable to quantify these satisfactorily, naturally factors of safety increased with time (*5*). The factor of safety had little scientific background, having an underlying empirical and subjective nature. No one can dispute that at the time that, say, stress analysis was in its infancy, this was the best knowledge available, but there was also the fact that there were a sufficient number of failures still happening to conclude that deterministic approach does not always ensure intrinsic reliability (*6*). A great disenchantment with factors of safety grew over the years (*7*).

As early as the 1920s, however, it was suggested that design performance should be based on means and variances of the random variables involved (*8*); a totally different mindset at the time. In the 1940s, a statistical basis to the factor of safety was proposed (*9, 10*). In the 1950s, engineers began to think differently about design, say, with the introduction of a 'true margin of safety', demonstrating that engineering problems are multi-factored, and variability based (*11*). With the increasing use of statistics in engineering around this time, and the advent of mass production, these ideas were the real enablers for probabilistic approaches, and in the 1960s, the theoretical developments governing probabilistic design, its paradigms, and methods were becoming established (*4, 12–14*). Many practical methods have since been developed to cope with the full range of engineering problem types, model formulations, and data characteristics (*15*).

In this paper, the basic underlying principles of the probabilistic approach are presented, and a comparison made to the deterministic approach, illuminating the former's benefits. A classification of the types of application, or utilities, of probabilistic methods is presented, followed by a number of design and structural integrity case studies which are used as exemplars for the specific application objectives: reliability prediction, performance assessment, optimisation, sensitivity analysis, and uncertainty characterisation. The paper closes with a discussion of the implementation issues surrounding probabilistic approach, before conclusions are presented.

II. The Basic Principles

Virtually all engineering parameters such as dimensions, material properties, and service loads exhibit some statistical variability that influence the adequacy of the problem, and therefore should be treated as random variables. The main engineering random variables that should be adequately described when using a probabilistic approach are shown in Fig. 1. The variables conveniently divide into two types: (*i*) design dependent, which the designer has the greatest control over, and (*ii*) service dependent, which the designer has limited control over.

Typically, the most important design-dependent variables are material strength and dimensional variability. Material strength can be statistically modelled from sample data for the property required; however, difficulties exist in the collation of information about the properties of interest. Dimensional variability and its effects on the stress acting on

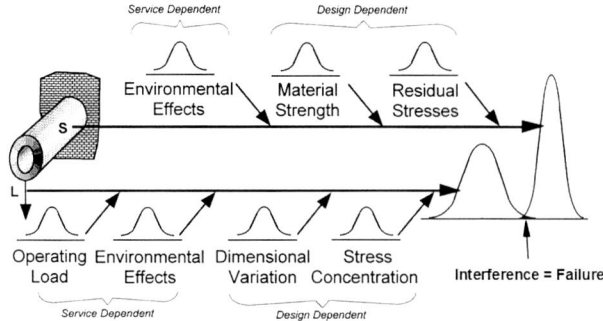

Fig. 1. — Key variables in a probabilistic design approach [Booker et al. (5)].

Table I

Typical coefficient of variation, CV, for a variety of engineering parameters [Booker et al. (5)].

Loads
Aerodynamic loads in aircraft = 0.012-0.04
Spring force = 0.02
Bolt preload using powered screwdrivers = 0.03
Aircraft thrust loads = 0.05
Thermal loads = 0.08
Powered wrench torque = 0.09
Dead load = 0.1
Hand wrench torque = 0.1
Vibration loads = 0.2
Live load = 0.25
Snow load = 0.26
Human arm strength = 0.3-0.4
Wind loads = 0.37
Acoustic loads in aircraft = 0.4
Transient loads = 0.5

Material Properties
Steels
Ultimate Tensile Strength = 0.05
Yield Strength = 0.05-0.2
Endurance Strength = 0.08
Brinnel Hardness = 0.05
Modulus of Elasticity = 0.01-0.03
Modulus of Rigidity = 0.02-0.04
Fracture Toughness = 0.05-0.1
Poisson's Ratio = 0.025

Geometry Parameters (produced by manufacturing processes)
Grinding = 0.00015
Turning/boring = 0.0004
Powder metal sintering = 0.0006
Drilling = 0.001
Milling = 0.003
Hot rolling = 0.008
Closed die forging = 0.009

Other Engineering Parameters
Coefficient of linear expansion for metals = 0.01
Stress concentration factor for machined notched bar = 0.03
Surface tensile residual stress for turning/boring = 0.1
Nut factor for cadmium plated bolt = 0.15
Surface roughness for turning/boring = 0.2
Coefficient of friction for steel on steel shrink-fit = 0.22

Other Materials
Ultimate Tensile Strength for cast iron = 0.09
Ultimate Tensile Strength for wrought iron = 0.04
Rupture strength for carbon fibre composites = 0.17
Modulus of Elasticity for cast iron = 0.04
Modulus of Elasticity for titanium = 0.09
Modulus of Elasticity for aluminium = 0.03
Shear and compression strength for honeycomb = 0.1
Tensile strength for honeycomb = 0.16
Ultimate tensile/yield strength for non-ferrous metals = 0.05

a component can be great, but information is typically lacking about its statistical nature, and its impact on geometric stress concentration values are rarely assessed. Important service-dependent variables are related to the loading of the component and stresses resulting from environmental effects. These are generally difficult to determine at the early stages of a project because of the cost of performing experimental data collection, the nature of overloading and abuse in service, and the lack of data about service loads. Also, the effect that service conditions have on the material properties is important, the most important considerations arising from extremes in temperature, as there is a tendency towards brittle fracture at low temperatures, creep rupture at high temperatures.

Several researchers and organisations over the last few years have accumulated statistical data for important engineering properties. Table 1 shows the scale of the variability of these parameters in terms of the coefficient of variation, CV (the standard deviation divided by the sample mean) (5). If variability is said to be a major source of unreliability in an engineered system or product (16), then observing the data in Table I suggests that the quality control of the manufacturing processes and material properties might not be as important as controlling the service environment and loads in achieving high reliability (6). The random nature of the engineering parameters and the scale of the variation in each is therefore well known. Engineers are familiar with the typical appearance of sets of strength data from tensile tests in which most of the data values congregate around the mid-range with decreasingly fewer values in the upper and lower tails on either side of the mean. For mathematical tractability, the experimental data can be modelled with a probability density function (PDF) or continuous distribution that will adequately describe the pattern of the data using just a single equation and its related parameters—normal, Weibull, log-normal, and extreme value types, to name just a few.

Composing these statistically characterised parameters in the performance model which reflects the objective of the problem, we must then use certain probabilistic methods to essentially simulate large numbers of random events taking place. This argument is extended when considering the probability of failure of a component, which is based on the joint probability of interference of the inherent material strength distribution (S), and loading stress distribution, (L), where both are random variables (see Fig. 1). When a random stress exceeds a random strength on any simulation event, failure is recorded, and this continues for all events until a probability of failure can be predicted accurately enough. In essence, the interference between the actual stress and strength distributions dictates the performance of the product in service and this is the basis of the probabilistic approach. Accurate representations of stress and strength as distributions, therefore, enables a meaningful failure prediction to be generated.

III. Why a Probabilistic Approach?

As outlined from the above arguments, a probabilistic approach incorporates the uncertainties with typical design inputs fully using any experimental data available, and thus provides the required realism when modelling a problem (1). It also provides quantitative measures of performance, helping engineers draw conclusions from complex analyses with a high degree of confidence. Increasing demands for higher performance and efficiency, resulting often in operation near limit conditions, has placed increasing emphasis on precision and realism in this way (17, 18). Increased use of analytical and simulation techniques in engineering is facilitating a move to fully simulation-based design, which for required confidence, must also be conducted probabilistically. This will eventually lead to cost savings on prototype validation prior to the production and installation of engineered products and systems. But probabilistic approaches are also useful where test to failure is not a practical proposition, where weight minimisation and (or) material cost reduction is important (i.e. not overdesigned), or where a safety case has to be made, e.g. in a component structural integrity assessment for operational plant. The only alternative is to resort to a deterministic approach using factors of safety, which can lead to either an unconservative design with unacceptable high failure rates, or a very conservative design that provides the required performance with unnecessarily high costs (19). Either way, determinism is conservatism, which is ineffective for the modern world (20). Table II summarises some of the important characteristics of probabilistic approach compared to a deterministic one.

IV. Case Studies

A number of important applications of probabilistic methods exist in reliability prediction, as discussed earlier, specifically where it would be useful to explore the level of random failure, resulting from the interaction of the distributions of loading stress and material strength. Other benefits from building a probabilistic model also present themselves, such as the purpose of a sensitivity analysis to provide a measure of the contribution of each of the input design- and service-dependent parameters to the problem. Only a few parameters have a significant impact on output variability generally (i.e. Pareto's rule analogy) and can lead to a better characterisation of the variability in the most influential parameters, with a focus on quality control and increased data collection and characterisation of the properties of interest and influence. This can also speed up a probabilistic assessment by negating trivial parameters in the problem formulation and computation. Sensitivity analysis can be snapshot (time independent) or time dependent, if the operating or environmental conditions change with time. Other utilities are to validate the output of a model or several models against experimental

TABLE II

Main utilities of a probabilistic approach.

Application	Objectives
Reliability/risk assessment	Evaluate the probability of failure (or reliability)
Performance assessment	Given a design, what range performances can I expect? Provide measures for improvement
Optimisation	Reduce redundancies through more economic design Optimise a design given multiple competing requirements
Sensitivity analysis	Which inputs contribute the most to the output uncertainty?
Characterisation of uncertainties	Increase confidence in processes and outputs

TABLE III

Deterministic vs probabilistic approach.

	Deterministic	Probablistic
Inputs	Single values, e.g. lower bounds	Distributions, histograms, ranges, or single values
Outputs	Pass/fail result	Probability of failure (single value)
Correlations	Based on judgement	Measured from data
Sensitivity study	Local	Local and global
Run time	Single runs	Typically $> 10^5$ runs

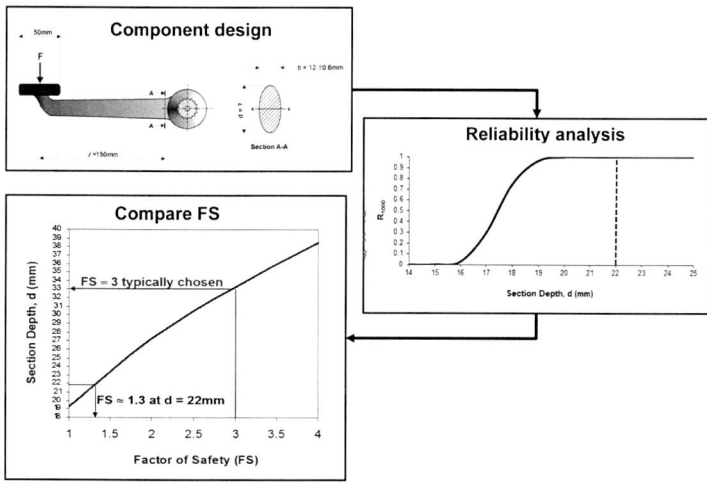

FIG. 2. — Reliability/risk analysis: component quasi-static design [Swift et al. (21)].

results with confidence measures and to provide optimisation of the parameters involved in the problem given some target, e.g. probability of failure. Table III shows the main utilities of a probabilistic approach.

Fig. 2 overviews the probabilistic analysis of a forged steel foot pedal under quasi-static loading (1,000 load applications), with the objective of finding the section depth for intrinsic reliability (21). The outcome suggests this is achieved at 22 mm section depth, compared to 33 mm from a deterministic approach using a factor of safety of three, typical for this type of problem. The potential for overdesign and excessive use of material and increased cost is evident using the deterministic approach. Within the topic

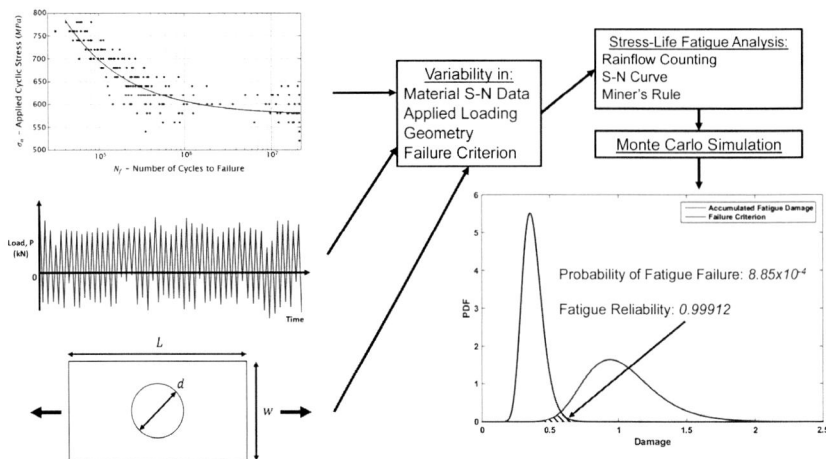

Fig. 3. — Reliability/risk analysis: component fatigue design [Socie. (24); IHS ESDU (25)].

of fatigue design, a number of parameters exhibit variability, including the scatter present within $S-N$ data, uncertainty within the loading spectrum and dimensional variability (22) (see Fig. 3). This leads to variability in the accumulated fatigue damage computed by Miner's rule which can be simulated using a Monte Carlo simulation. As the failure criterion for Miner's rule also exhibits variability (23), a stress–strength interference approach can be used to compute the probability of failure of the component, to support design decisions. Fig. 4 presents the probabilistic assessment of three dif-

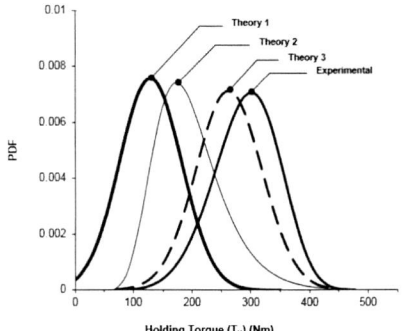

Fig. 4. — Performance assessment: shrink-fit design [Booker et al. (26)].

ferent theories for shrink-fit holding torque compared to the experimental distribution composed from 27 samples; the preferred theory being 3, which is 'closer' to the experimental, but is statistically verified as being the same distribution at a certain confidence level (26).

Similar to the previous case in its objective, Fig. 5 shows the experimental creep damage results for a specimen under uniaxial loading compared with two competing probabilistic models for creep relaxation. A greater degree of confidence is related to the use of 'Model 2' compared to 'Model 1' (27). The solenoid design shown in Fig. 6 has two failure modes: (i) failure potential at the weakest section by stress rupture due to the assembly torque, and (ii) that the pre-load on the solenoid thread section is sufficient to avoid loosening in service. It is necessary to determine the mean assembly torque, M, to

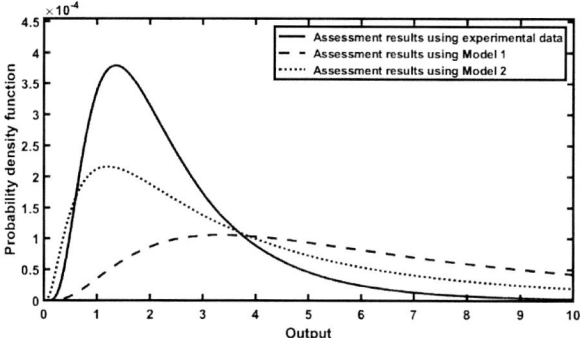

FIG. 5. — Performance assessment: creep damage [Zentuti et al. (27)].

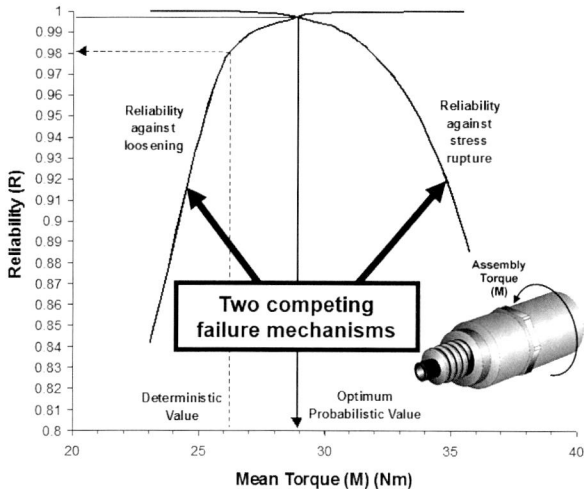

FIG. 6. — Optimisation: competing failure modes [Booker et al. (5)].

satisfy these two competing failure modes using a probabilistic design approach, where the target torque to be applied is an optimum at the highest reliability achievable. Fig. 7 shows the sensitivity analysis results from the relative contribution of six input parameters, which are all described as random variables, towards the output of a creep–fatigue damage assessment, using four different sensitivity analysis methods (27). Using four different methods (local, global) allows corroboration of the results, although, the main interest is gauging which parameters require further data collection, quality control, etc. Fig. 8 shows a time-dependent sensitivity analysis and the shift in the relative importance of two input parameters owning to the time-dependent nature of creep. Fig. 9 provides the probabilistic creep–fatigue damage results as compared with the outcome of a conventional deterministic calculation of the same type, showing the relative conserva-

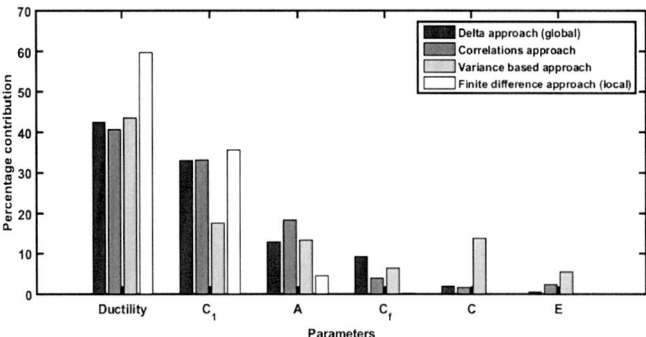

Fig. 7. — Sensitivity analysis: uniaxial creep–fatigue assessment [Zentuti et al. (27)].

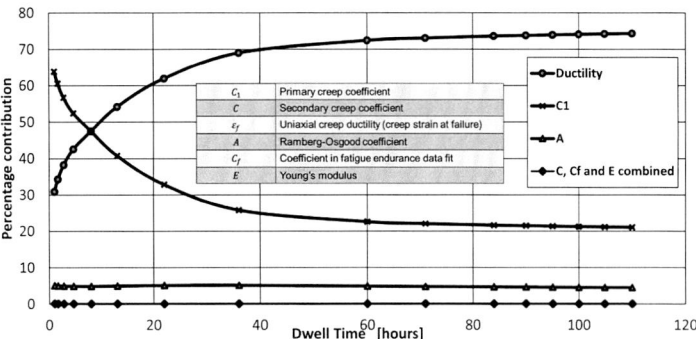

Fig. 8. — Time-dependent sensitivity analysis: uniaxial creep–fatigue assessment.

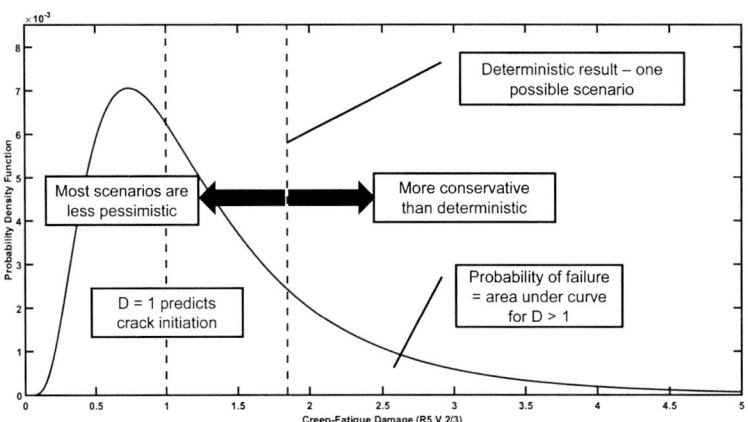

Fig. 9. — Uncertainty characterisation: creep–fatigue damage.

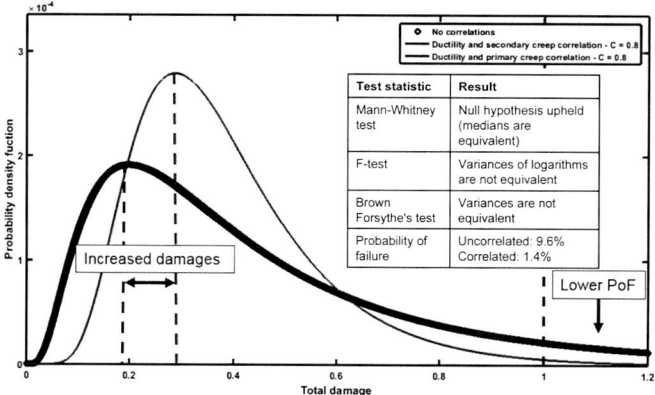

FIG. 10. — Effect of correlations on creep–fatigue damage.

the development cycle you are in. The robustness of the chosen probabilistic methods towards the type of system or problem to be modelled is also an important selection decision, i.e. completeness and statistical description of data used, engineering model type (analytical, numerical, empirical, non-linear), and objective to be achieved—optimisation, probability of failure prediction, and so on (15). For example, the evaluation of many engineering systems requires the use of numerical methods such as finite element analysis. Therefore, numerically efficient algorithms are required for probabilistic approaches when applied in order that time and costs are reduced (29).

V. Conclusions

The measures of performance determined using a probabilistic approach give the designers more confidence in their assessments, analyses, and predications by providing better understanding of the variables involved and quantitative estimates for failure probability. Although associated methods have been adopted in some companies and have been successfully applied to specific cases, a cultural and educational step-change is necessary for probabilistic approaches to be integrated in mainstream engineering activities leading to a probabilistic 'mindset'. A lack of awareness of their existence or usefulness, computational intensity, requirement for statistical knowledge, difficulty and the cost of data collection, are possibly a few of the obstacles that may have limited their use to date. Industry also often finds it difficult to justify the resource and training commitments needed to support these activities, and a key problem has been in consolidating the knowledge for advancing their utility. Probabilistic design is seen as an opportunity for companies to enhance their competitive advantage through optimisation, reliability, and performance target qualification. It provides a transparent means of explaining to a business more about the safety aspects of engineering design decisions with a degree of clarity not provided by the 'factor of safety approach'. Although significant advances have been made in probabilistic methods in terms of efficiency and accuracy, their integration in industrial product development is unsatisfactory. The focus now should now be on the routine use of probabilistic approaches by companies across all sectors. It is hoped that new and existing practitioners alike will gain some confidence to progress with probabilistic applications from the examples presented which span a range of different structural integrity and design analysis problems.

Acknowledgements and Dedication

The work reported in this paper has been supported by TRW Automotive Electronics, EDF Energy and Safran Landing Systems. This paper is dedicated to the late Professor John Knott of the University of Birmingham in recognition of his contributions to the fields of structural integrity and materials science.

References

1. K. V. Bury, "Statistical Models in Applied Science". Wiley, New York, NY, USA, 1975.
2. S. F. Wojtkiewicz, M. S. Eldred, J. R. Field, A. Urbina, and J. R. Red-Horse, *in* "Proceedings" 42nd AIAA/ASME/ASCE/AHS/ASC Structures, Structural Dynamics, and Materials (SDM) Conference, Seattle, Washington, USA, 16th–19th April 2001. Paper No. AIAA-2001-1455.
3. J. Clausen, S. O. Hansson, and F. Nilsson, *Reliab. Eng. Syst. Safe.* **91,** 964 (2006).
4. H. O. Madsen, S. Krenk, and N. C. Lind, "Methods of Structural Safety". Dover, New York, NY, USA, 2006.
5. J. D. Booker, M. Raines, and K. G. Swift, "Designing Capable and Reliable Products". Butterworth-Heinemann, Oxford, UK, 2001.
6. A. D. S. Carter, Mechanical Reliability." 2nd Edition. Macmillan, London, 1986.
7. G. E. Dieter, "Engineering Design: a Materials and Processing Approach." First Metric Edition. McGraw-Hill, New York, NY, USA, 1986.
8. M. Mayer, "Die Sicherheit der Bauwerke". Springer Verlag, Berlin, Germany, 1926.
9. A. G. Pugsley, *Aircraft Eng. Aero. Tech.* **16,** 18 (1944).
10. A. M. Freudenthal, *Trans. Am. Soc. Civil Eng.* **112,** 125 (1947).
11. F. Nixon, *in* "Proceedings" Conference on Technology of Engineering Manufacture, Institution of Mechanical Engineers, London, UK, 14th March 1958, 561.
12. J. H. Saleh and K. Marais, *Reliab. Eng. Syst. Safety* **91,** 249 (2006).
13. C. Lipson, N. Sheth, and R. L. Disney, "Reliability Prediction—Mechanical Stess/Strength Interference", Final Report No. RADC-TR-66-710. Rome Air Development Center, Griffiss Air Force Base, New York, NY, USA, 1967.
14. E. B. Haugen, "Probabilistic Approaches to Design." Wiley, New York, NY, USA, 1968.
15. Y. M. Goh, C. A. McMahon, and J. D. Booker, *Proc. Inst. Mech. Eng. O* **223,** 199 (2009).
16. A. D. S. Carter, "Mechanical Reliability and Design." Macmillan, London, UK, 1997.
17. E. B. Haugen, "Probabilistic Mechanical Design." Wiley-Interscience, New York, NY, USA, 1980.
18. D. J. Smith, "Reliability, Maintainability and Risk: Practical Methods for Engineers." 4th Edition. Butterworth-Heinemann, Oxford, UK, 1993.
19. R. C. Rice, *in* "ASM Handbook No. 20, Materials Selection and Design." 10th Edition. ASM International, Novelty, OH, USA, 1997.
20. M. Modarres, "What Every Engineer Should Know about Reliability and Risk Analysis." Marcel Dekker, New York, NY, USA, 1993.

21. K. G. Swift, M. Raines and J. D. Booker, *J. Eng. Design* **11,** 299 (2000).
22. J. Hoole, P. Sartor, and J. Cooper, *in* "Proceedings" Royal Aeronautical Society 5th Aircraft Structural Design Conference, Manchester, UK, 4th–6th October 2016.
23. P. H. Wirsching, *in* "Probabilistic Structural Mechanics Handbook: Theory and Industrial Applications", (C. Sundararajan, ed.), pp. 146–65. Chapman & Hall, New York, NY, USA, 1995.
24. D. Socie, "eFatigue."
   http://www.efatigue.com (accessed 21st March 2018)
25. IHS ESDU, "Data Item 04019 (A). Endurance of High-Strength Steels." IHS Engineering Sciences Data Unit, London, UK, 2006.
26. J. D. Booker, C. E. Truman, S. Wittig, and Z. Mohammed, *Proc. Inst. Mech. Eng. B* **218,** 175 (2004).
27. N. A. Zentuti, J. D. Booker, R. A. W. Bradford, and C. E. Truman, *Mater. High Temp.* **34,** 333 (2017).
28. J. E. Shigley and C. R. Mischke, "Mechanical Engineering Design." 5th Edition. McGraw-Hill, New York, NY, USA, 1989.
29. H. A. Jensen and A. E. Sepulveda, *AIAA J.* **38,** 2133 (2000).

Generic Design Assessment and Structural Integrity Challenges: Past, Present, and Future?

J. P. Caul

ABSTRACT.—New nuclear build is a UK Government priority. The Office of Nuclear Regulation (ONR) is undertaking generic design assessment (GDA) of prospective designs of new nuclear power plants for the UK. ONR introduced GDA as a pre-authorisation process for new reactor designs. GDA is the start of the life cycle for a new reactor and is a comprehensive assessment of the safety, security and environmental aspects of the design in advance of any construction on site. It is undertaken jointly by ONR, the Environment Agency and Natural Resources Wales. GDA aims to de-risk construction by providing regulatory clarity on design changes and safety analysis ahead of financial decisions. GDA is a multistep process with increasing levels of scrutiny based on a claims, arguments, and evidence structure. GDA is an open and transparent process. GDA was developed on the premise of 'large' established or available reactor designs. ONR is looking at options to develop a more flexible design assessment process (and seeking to improve efficiency), which remains consistent with previous GDAs, achieves the same objectives, and provides an enabling approach to SMR development and deployment.

This paper provides an overview of the GDA process and discusses some challenges from a structural integrity perspective. It includes reviews of past GDAs, outlines the status of the reactor designs within GDA, and identifies some future challenges relating to the structural integrity assessment for advanced nuclear technologies, i.e. small modular reactors (SMRs) and advanced modular reactors (AMR). These relate to meeting UK regulatory expectations, which are informed by precedent and the UK's non-prescriptive regulatory regime.

I. Introduction

In the UK, the advanced gas-cooled reactor (AGR) civil nuclear power reactors are moving towards the end of their operating lives. The UK Government has decided it wants nuclear power to be one of the options for future electricity generation. The Office for Nuclear Regulation (ONR) is the UK's statutory nuclear regulator and licenses any new civil nuclear power plants in the UK. The ONR is an independent statutory corporation bringing together the safety and security functions of the Health and Safety Executive's (HSE) former Nuclear Directorate, including Civil Nuclear Security and UK Safeguards Office and Radioactive Material Transport from the Department for Transport. ONR's role is captured in its mission statement:

> "To provide efficient and effective regulation of the nuclear industry, holding it to account on behalf of the public."

Principal Inspector, Office for Nuclear Regulation (ONR), Bootle, Merseyside L20 7HS, UK.

ONR is undertaking an assessment of prospective designs for new civil nuclear power plants for the UK. In 2007, the Nuclear Directorate (ND) of the Health and Safety Executive (now ONR), introduced a 'new' approach, the GDA process, to reviewing and licensing the construction of new nuclear power plant designs. The GDA process is explained by ONR (*1*), Caul and Holt (*2*), and Harrop (*3*). There are three parts to the regulation of new reactors:

Part 1: Generic Design Assessment (GDA). — In four steps.
- safety assessment of the nuclear power plant design
- use of generic site characteristics

Part 2: Nuclear Site Licensing and Authorisation.
- site specific
- operator specific

Part 3: Construction.
- permissioning (assessment and inspection)
- compliance and conformity with licence conditions, e.g. LC 19 and LC 20

This paper provides an overview of the GDA process, reviews past GDAs, outlines the status of the reactor designs within GDA and outlines potential developments in the design review process. These developments reflect the learning and experience gained from past GDAs whilst also anticipating the future challenges associated with the introduction of advanced nuclear technologies (ANT), i.e. light or heavy water cooled small modular reactors (SMR) and advanced modular reactors (AMR). The ONR definition of structural integrity covers the integrity of metal structures and components. These primarily are metal pressure boundary components but they also include metal support structures and free standing metal containment structures. This paper identifies the key challenges for structural integrity including some expected challenges associated with the introduction of ANT. These relate to meeting UK regulatory expectations, which are informed by precedent and the UK's non-prescriptive regulatory regime.

II. Regulation of New Nuclear Build and the GDA Process

1. ONR's Regulatory Philosophy and Regulation of New Build

The safety of nuclear installations in Great Britain (GB) is assured by a system of regulatory control based on a licensing process by which a corporate body is granted a licence to use a defined site for specified activities. ONR was established by the *Energy Act 2013* (EA13). ONR's principal function is to take such action as it considers appropriate for the purposes for which it was established. These purposes relate to nuclear safety, conventional health and safety on nuclear sites, nuclear security, nuclear safeguards, and the transport of radioactive materials. UK nuclear safety law, in common with UK conventional safety law, is non-prescriptive and goal setting. In line with this framework, ONR does not licence technologies for deployment in the UK, ONR permissions activities, through a process of specifications, agreements, consents, directions, approvals, and notifications. These are exercised via conditions attached to the licence. A more detailed overview of the licensing of new nuclear installations is described by Hopkin (*4*).

ONR's approach is founded on developing and sustaining an open and effective dialogue with duty holders via the adoption of an enabling approach whilst ensuring the duty holders compliance with legal duties. The overall regulatory process for new nuclear build comprises three elements, namely: (*i*) GDA; (*ii*) nuclear site licensing; and (*iii*) construction. The focus of this paper is the GDA process in the context of structural integrity assessment.

2. GDA Process

In 2006 the Government carried out an energy review concluding that nuclear power would have a role in the future UK generating mix. The review proposed a number of initiatives to reduce the regulatory barriers for new nuclear build. The Government asked HSE and the other principal nuclear regulators (the Environment Agency, the Scottish Environment Protection Agency, and the Office for Civil Nuclear Security) to implement a 'pre-authorisation' system for reactor designs to allow generic designs to be assessed in advance of any application to build a nuclear power station at a particular location.

In response ONR developed the GDA process. GDA is the start of the life cycle for a new reactor and is a comprehensive assessment of the safety, security, and environmental aspects of the design in advance of any construction on site.

GDA facilitates early regulatory engagement and in so doing maximises the opportunity for regulatory influence. GDA aims to de-risk construction by providing regulatory clarity on design changes and safety analysis ahead of financial decisions. This allows for the identification and resolutions of key issues to reduce the cost and time risks long before build. GDA also maximises the value of pre-application by simplifying the site specific phase. GDA is undertaken by ONR, the Environment Agency (EA), and Natural Resources Wales (NRW). This enables a joined-up regulatory approach with regulators working together to clarify expectations and to assure consistency. The GDA is an open and transparent process; for example, to build public confidence all regulator reports have been published from the outset.

The GDA process has been in place for over ten years and is well established. There are currently four steps which provide the basis for increasing levels of regulatory scrutiny. The GDA process and steps are shown in Fig. 1 and detailed in ONR (*1*).

The GDA steps are as follows:

Step 1; preparatory phase. Key objective is for the requesting party (RP) to develop adequate project management arrangements and deploy sufficient technical resource to complete the GDA. In the absence of a license, RP is the term used by ONR for those organisations that request their designs to be assessed.

Step 2; a high-level technical assessment of the fundamental aspects of the design (claims). Key objective is to identify any fundamental safety or security issues that might

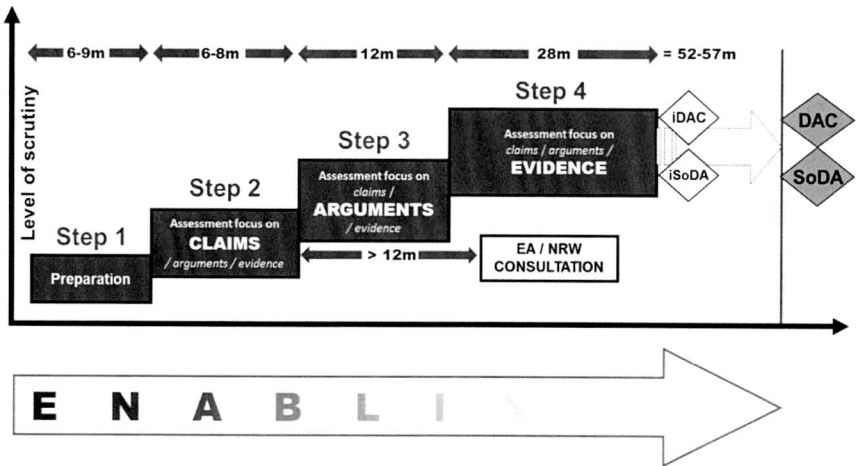

DAC: ONR's Design Acceptance Confirmation (iDAC: interim DAC)
SODA: EA/NRW's Statement of Design Acceptability (iSODA: interim SODA)

FIG. 1. — An overview of the generic design assessment (GDA) process.

prevent the issue by ONR of a design acceptance confirmation (DAC) or a statement of design acceptability (SODA) by the EA and NRW.

Step 3; a more detailed assessment of the design (arguments). Key objective is to identify whether any significant design changes are required to meet UK expectations

Step 4; an in-depth assessment of evidence presented by the RP to make the safety case for the design. Key objective is for the regulators to complete a detailed assessment to enable a judgement of whether a DAC should be issued.

Assessment at Step 3 and Step 4 may use a combination of internal technical expertise and technical support contractors (TSC). However, where TSC support is engaged, the regulatory judgements are made by ONR.

On completion of GDA, ONR issues a DAC. The DAC remains valid for a period of ten years from the date of issue. Similarly, completion of the environmental assessment is marked by the EA issuing a SODA. Note that if at the end of Step 4, the ONR or EA/NRW are only partially content, then an interim acceptance is issued through an iDAC/iSODA. This results in GDA Issues and the development of RP resolution plans to close-out. The RP has to successfully implement and complete resolution plans in order to get the DAC/SODA.

GDA is not a formal regulatory legislative requirement. The DAC does not guarantee that ONR will grant permission for the construction of a power station at a particular site in Great Britain. Any organisation wishing to build and operate such a nuclear installation in Great Britain must obtain from ONR a site-specific licence under the *Nuclear Installations Act* 1965 and any necessary consents for construction of that installation in accordance with the conditions attached to that licence. However, ONR will, during the period for which it remains valid, take the DAC into consideration in assessing the adequacy of any licensee's case for requesting consent for the start of nuclear island safety related construction for a site licensed in Great Britain.

One outcome from GDA, following provision of a DAC, is a commitment from ONR not to further assess at the site-specific stage of the project aspects of the safety and security case already assessed and accepted at the generic design stage.

A reactor design which is acceptable to other regulatory authorities across the world may not be acceptable in the UK. This is because of the different regulatory regimes and phasing approaches. For example, the US uses a rule-based design certification approach which is not equivalent to GDA. There are also different regulatory expectations for the scope and content of safety cases along with responsibilities for safety. In addition, under the *Health and Safety at Work Act*, in the UK, there is an expectation that risks are reduced 'so far as is reasonably practicable' (SFAIRP) which leads to differing expectations. Note that SFAIRP is effectively equivalent to 'as low as reasonably practicable' (ALARP), but SFAIRP is the legal term.

For each reactor assessment, ONR identifies technical questions and issues. These are managed through interface arrangements, with a tiered approach as follows:

a. Regulatory Query (RQ). These are requests by ONR for clarification and additional information and are not necessarily indicative of any perceived shortfall.

b. Regulatory Observation (RO). An RO is raised when ONR identifies a potential regulatory shortfall and requires action and new work for it to be addressed. Each RO can have several associated actions. ROs are published on the joint regulators' website.

c. Regulatory Issue (RI). An RI is raised when ONR identifies a serious regulatory shortfall which is potentially significant enough to prevent provision of a DAC, and requires action and new work for it to be addressed. Each RI can have several associated actions. RIs are published on the joint regulators' website.

It is possible that a question raised as an RQ could escalate to an RO or to an RI. In addition, ONR may issue ROs as a means of providing advice to the RP to inform the development of their plans and programme, i.e. to expedite progression of the GDA.

The fact that ONR has identified these issues should not in all cases be interpreted as us being overly critical of the reactor designs; rather they should be seen as evidence of an independent and robust regulatory process. It is also evidence that GDA is working as intended and allowing us to have influence on the design and safety case well in advance of construction in the UK.

GDA is designed to assess the generic safety case for future reactor designs; it is not intended to provide a complete assessment of the final reactor design, as there will be other factors, operator specific or site related, that ONR would expect to consider during the site specific stages. Also, in some instances, final validation of the safety case can only be completed when the final detailed design of equipment is developed by a manufacturer or supplier, or when the facility is being constructed and is in the process of being tested. This validation process is normal regulatory business and will be subject to appropriate regulatory controls. The link from GDA to site-specific assessment is assured by the identification of GDA assessment findings. These findings are primarily concerned with the provision of additional safety case evidence, after GDA, to confirm certain safety aspects as the project progresses through the detailed design, construction

3. GDA Current Status: Structural Integrity Perspective

The EDF and AREVA UK EPR™ (a pressurised water reactor [PWR]) along with the Westinghouse AP1000®, (a PWR with passive safety features) and the Hitachi GE UK ABWR (an advanced boiling water reactor) have completed GDA. The position of the assessment of the reactor designs in relation to the GDA is outlined below.

For the UK EPR™, an iDAC and iSODA were issued at the end of Step 4 in December 2011. A total of 31 GDA issues were raised across the disciplines with two relating to structural integrity: avoidance of fracture, and the reactor pressure vessel (RPV) surveillance scheme. All issues were closed out, which culminated in the issuing of a DAC and SODA in December 2012, the first such confirmation granted by ONR. Note that 82 design changes were identified and there were approximately 600 assessment findings for the future licensee to address during licensing and construction (5).

The Westinghouse AP1000® reactor completed Step 4 with the issue of an iDAC and iSODA in December 2011. Westinghouse Electric Company (WEC) made a decision not to close out the GDA so this was followed by a pause. There were 51 outstanding GDA issues and approximately 580 assessment findings across the disciplines for the future licensee. There were six issues relating to the structural integrity discipline. These covered: (*i*) avoidance of fracture; (*ii*) fatigue analyses of ASME III Class 1 pipework; (*iii*) reactor coolant pump safety case and flywheel disintegration case; (*iv*) fracture analysis of the containment vessel; (*v*) compliance with ASME III design rules; and (*vi*) structural integrity related classification topics. WEC remobilised to close out the GDA in September 2014 with a target completion date of early 2017. Resolution of the six GDA structural integrity issues was completed in March 2017 to support the issue of a DAC (6).

The Hitachi-GE UK ABWR completed Step 2 and Step 3 in 2014 and 2015 respectively. The GDA target completion date was December 2017. From a structural integrity perspective several ROs were issued at and post Step 2. These included: (*i*) avoidance of fracture; (*ii*) control rod drive (CRD) penetration design; (*iii*) RPV design; (*iv*) RPV material specifications; (*v*) material selection; and (*vi*) reinforced concrete containment vessel (RCCV) drywell head integrity. Resolution of the ROs was completed in GDA Step 4, supporting a DAC in December 2017. The DAC included 201 assessment findings, 21 of which were raised by ONR's structural integrity discipline (7).

The UK HPR1000 (General Nuclear Systems) is a PWR design which commenced Step 2 of GDA in November 2016. Step 2 is due to complete in June 2018 with issue of ONR's assessment report in November 2018. In common with previous GDAs, from a structural integrity perspective key areas of interest include: (*i*) the overall approach to structural integrity demonstration; (*ii*) categorisation and classification; (*iii*) applicable codes and standards; (*iv*) safety case strategy, avoidance of fracture, design summaries for major vessels and piping; (*v*) material selection principles along with ageing and degradation mechanisms; and (*vi*) ALARP.

4. ONR Review of Generic Design Assessment Process

The GDA process has been completed for three reactor designs to date (the EDF AREVA UK EPR™, Westinghouse AP1000®, and Hitachi GE UK ABWR) with a fourth GDA in progress at the GNS UK HPR1000. Valuable lessons have been learnt from undertaking these GDAs which, as part of continuous improvement, ought to be reflected in the process. Many of these lessons relate to improving regulatory efficiency. It is therefore an appropriate time to review the GDA process more generally. Nevertheless, the underlying objectives for GDA remain valid, i.e. to reduce regulatory risks.

GDA was originally developed on the premise of 'large' commercially available reactor designs being deployed at multiple sites. These designs have, by definition, achieved a certain level of design maturity and have established safety cases which may have been reviewed or approved by other regulatory bodies. However, there is a growing interest in the potential deployment of ANT: SMRs (water cooled) and AMR (Generation IV-type) in the UK. For SMRs, the regulatory landscape is changing and so the underlying premise for the existing GDA process may not be suitable. There are also other important differences which need to considered, for example relating to the maturity of the technology, novelty of features, and safety and security case approaches, which may require evolution of the design assessment process. Thus, as part of continuous improvement, the regulators are looking at whether there are elements of the GDA process that could be modernised to:

- improve flexibility and better adapt to the differing levels of maturity and development of SMR vendors and their technologies
- capture important lessons learnt from previous and ongoing GDAs
- remain consistent with previous GDAs

ONR is therefore currently considering options to develop a more flexible design assessment process (and seeking to improve efficiency) which remains consistent with previous GDAs, achieves the same objectives, can accommodate SMR technology and requirements, and provides an enabling approach to SMR development and deployment. In short, ONR's approach to the review of the GDA process is evolutionary rather than revolutionary. The conclusions and recommendations of ONR's review of the GDA process are currently under due process, with publication expected before April 2019.

III. Structural Integrity Assessment

1. ONR Assessment Overview

Assessment is based on pre-construction safety reports submitted by the RP. The fundamental approach used in ONR's assessment is to examine the claims made in the safety case and seek arguments and evidence that support those claims. The assessment is not a comparison of designs but a review of each design against relevant guidance.

ONR Nuclear inspectors conducting assessment have as guidance the safety assessment principles (SAPs) (8) and technical assessment guides (TAGs). For structural integrity, key SAPs include those covering the integrity of metal structures and components (EMC.1 to EMC.34). Other SAPs which often inform the structural integrity assessment include: (*i*) safety classification and standards (ECS.1 to ECS.5); (*ii*) maintenance, inspection and testing (EMT.1 to EMT.8); (*iii*) ageing and degradation (EAD.1 to EAD.5); and (*iv*) pressure systems (EPS.1 to EPS.5).

The SAPs are underpinned by a suite of supporting technical assessment guides—the foremost one for structural integrity is NS-TAST 016 (*9*). Other TAGs that frequently inform the structural integrity assessment include "Guidance on the Demonstration of ALARP" (NS-TAST-GD-005) (*10*), "The Purpose, Scope and Content of Nuclear Safety Cases" (NS-TAST-GD-051) (*11*), and "Categorisation of Safety Functions & Classification of SSC" (NS-TAST-GD-094) (*12*).

The SAPs and TAGs give guidance to ONR inspectors in conducting their assessments. They require judgement in their use. They are not a fixed set of 'rules' for compliance. The SAPs and TAG 016 (*9*) cover a range of integrity situations from highest reliability (sometimes referred to as 'incredibility of failure' in the UK) down to 'non-nuclear' safety. The current SAPs are an evolution from earlier versions. A significant update of the SAPs followed the Sizewell B public inquiry (report: December 1986). The approach applied in the UK nuclear industry to date for highest reliability claims, arguments and evidence are broadly consistent with the outcome of the Sizewell B public inquiry. The SAPs and TAGs are used for all assessments, including sites that are currently operating and undergoing decommissioning, so they are not confined to 'new build' assessments.

The starting point for structural integrity assessment is the question "What is the hazard posed by failure (the consequences)?" This includes consideration of both the direct and indirect consequences of a postulated gross failure. For example, gross failure of a pressure boundary component could lead directly to release of radioactivity from the failed component, and (or) the event could be an 'internal hazard' that threatens the integrity of other components.

The assessment looks for measures taken to underpin structural integrity of a component that are consistent with the potential consequences of failure. If there are no engineered means of preventing or protecting against the consequences of a postulated gross failure, the safety case rests on avoiding the occurrence of the initiating event (the gross failure of a pressure boundary component). When the radiological consequences could be high, the likelihood of the initiating event needs to be correspondingly low. For metal pressure boundary components operating below the creep temperature region, the main potential failure modes are:

(1.) Rupture of the pressure boundary wall due to a combination of thickness and material strength not being sufficient to meet the loading demand.

(2.) Failure by propagation of a crack-like defect.

Pressure vessel and piping design and manufacturing codes have for many years dealt with failure mode 1 above. Pressure vessel and piping design and manufacturing codes also deal to an extent with failure mode 2 above. However, it is this second failure mode that receives closest attention in assessment of those components that are required to have the highest integrity.

2. Key GDA Structural Integrity Challenges

There are many challenges within GDA that are common to all disciplines. These include delivery of submissions and their assessment to tight programme timescales. ONR's resources are limited and there are other important regulatory programmes, including those for Sellafield and the operating nuclear facilities, where ONR has commitments to achieve regulatory outcomes.

There are also many technical challenges to address within the GDA. The following examples cover some key challenges from a structural integrity perspective. This is not an exhaustive listing. Rather, the examples serve to illustrate some common difficulties encountered in structural integrity assessment.

3. Reducing Risks ALARP

The GDA process is undertaken within the existing nuclear regulatory framework for Great Britain. The main element of this is the *Nuclear Installations Act* 1965, which sets down the requirement to obtain a nuclear site licence from ONR before installing a nuclear reactor on a site. It is underpinned by the more general *Health and Safety at Work Act* 1974, which places a basic responsibility on duty holders to reduce risk SFAIRP.

In the UK the nuclear safety regulatory regime is non-prescriptive. A corollary is that to comply with UK law duty holders are challenged to reduce risks to workers and the public SFAIRP. In terms of design, duty-holders need to provide evidence to demonstrate that a chosen design or concept reduces risks ALARP. This is particularly important at the design stage, which covers concept selection through to detailed design specification (drawings, calculations, specifications, and so on), because there is the maximum potential for reducing risks, by application of the principles of inherently safer design. The key questions are:

(1.) What can you do to reduce risk?

(2.) Why is it not reasonable to implement options & measures to reduce risk?

The answers involve balancing the benefits and detriments and the consideration of gross disproportion. These balances may be inclusive to structural integrity, but could also include other technical disciplines, for example material selection is important to the reactor chemistry and radiation protection disciplines.

In most cases demonstrating that risks are ALARP is not done through explicit comparisons of costs and benefits. Instead, ONR's judgements on the RP's ALARP justification are based on relevant good practice (RGP). RGP are those standards for controlling the risk judged and recognised by ONR as satisfying the law, when applied appropriately. For structural integrity RGP includes, for example:

(1.) Approved codes of practice, British, ISO, EC, IAEA standards, WENRA SRLs, ASME, RCC-M, SAPs, TAGs, etc.

(2.) What we have accepted previously in similar circumstances (R6, RSE-M, fracture assessments), R3 (impact procedure, previous GDAs, etc.)

(3.) The RP may propose what they consider RGP (may not be listed above), but ultimately ONR decides what is RGP.

ONR assessment involves benchmarking against RGP. The higher the risk or safety classification, the more rigorous the safety case and scrutiny; hence demonstrating that risks are ALARP are particularly important for highest reliability claims because the safety case is founded entirely upon the provisions for structural integrity.

ONR assessment is to judge compliance with these legal duties. ONR's intervention is proportionate and risk-based. ALARP considerations underlie several areas of interest within the structural integrity discipline, for example classification, material specification and selection, highest reliability structures and component design details, and design for 'inspectability'.

The demonstration that designs reduce risks ALARP has proven to be a challenging concept to RPs who are more familiar with prescriptive regulatory regimes (in the US and Japan, for instance). In contrast in the UK compliance with the provisions of an established nuclear design and construction code may just be the starting point to support a highest reliability claim. To assist duty holders in meeting their legal obligations, ONR provides advice and guidance. However, ultimately under UK law, duty holders are responsible for safety. In this respect, UK supporting organisations, who are more familiar with the UK regulatory regime and expectations also have obligations to provide sound advice to RPs.

4. Highest Reliability Components and Structures

There is a challenge to structure the pre-construction safety reports (PCSRs) in the form of a claims, argument, and evidence structure. This is particularly important for the structural integrity case when the RP needs to develop a case for 'highest reliability components and structures', i.e. where the likelihood of gross failure has to be inferred that it is so low that it may be discounted. SAP paragraph 286 indicates that discounting gross failure of a component or structure is an onerous route to constructing a safety case and that such a case should provide in-depth explanation of the measures over and above normal practice that support and justify the claim.

Key SAPs relevant to the development of a case for highest reliability include EMC.1 through EMC.34. Indeed, EMC.1 to EMC.3 are particularly important. EMC.1 includes the expectation that the safety case is robust and the corresponding assessment suitably demanding, in order to properly inform engineering judgement that:
- the metal component or structure is defect-free as possible
- the metal component or structure is tolerant of defects

EMC.2 includes the expectation of a comprehensive safety case and assessment including a comprehensive examination of relevant scientific and technical issues, taking account of precedent when available. Notable precedents include the UK legacy of the Lord Marshall and Sir Peter Hirsch reports and the Sizewell B public enquiry. In addition, previous GDAs may provide a good source of relevant good practice for RPs and the ONR.

EMC.3 expands on the evidence to support EMC.1 and EMC.2 with the caveat that the extent of the evidence should be commensurate with the importance to the overall safety case, which is informed by classification.

The development of a structural integrity case for highest reliability components and structures is relatively straightforward for UK-based licensees who are well aware of precedent and expectations, but this presents quite a challenge for non-UK based organisations.

A key point is that to underpin a highest reliability claim for a metal component or structure is that compliance with the provisions of an established nuclear design and construction code is in itself insufficient. ONR therefore sets expectations on what is needed over and above compliance with an established nuclear design and construction code. Examples of areas of interest (not exhaustive) include:

- identification of highest reliability components (HRCs)
- avoidance of fracture demonstration for HRCs
- defect tolerance calculations (R6 and RSE-M)
- qualified manufacturing inspection
- material properties, particularly fracture toughness
- material selection and specification
- compositional and forging processes for the main ferritic forgings, including the RPV
- design codes, ASME III and RCC-M
- irradiation damage and RPV material surveillance or testing
- documentary envelope and design reports
- environmental effect on fatigue design curves
- design for 'inspectability' including access requirements for in-service inspection (ISI)

Perhaps the most significant of the above is the avoidance of fracture demonstration. This is because, although this approach is informed by precedents in the UK, it is not adopted elsewhere. The approach integrates defect tolerance calculations for limiting defect sizes, and predictions of fatigue crack growth, with qualified inspections (examination) and material properties, notably, fracture toughness.

A 'margin' can be expressed as the ratio of the limiting defect size, corrected for fatigue crack growth, to the size of defect that can be found with high confidence by the manufacturing inspection. This approach implies 'qualification' of the weld inspections using, for example, the type of approach and principles established by the European network on inspection qualification (ENIQ). This is necessary because code inspections may not provide adequate assurance of the absence of defects of structural concern as derived from the defect tolerance calculations. For the GDA the emphasis is placed on gaining assurance that UK expectations are understood, by focussing efforts on the most limiting areas, usually the welds, in highest reliability components. Ultrasonic inspection approaches are often required to supplement radiography. In addition, the capability for other manufacturing inspections including forging inspections is assessed. The scope of the evaluation is extended to a broader consideration of regions along with the ISI as part of licensing and construction. However, as a precursor to the development of the licensees ISI provisions, the need for 'inspectability' including access and design provisions are considered in GDA, for example, in the UK EPR™ the main coolant loop crossover leg redesign to improve 'inspectability'.

The R6 defect assessment procedure is acknowledged as being a suitable approach for undertaking defect tolerance calculations in the UK. However, EDF and AREVA chose to use the French-based RSE-M methodology. ONR is not prescriptive, and provided that the approach has been suitably validated and verified (SAP EMC.34), then it could be acceptable. ONR commissioned a TSC to undertake comparative calculations using the same input data and found that the plasticity correction factor used to account for the interaction of primary and secondary stresses appears to be different. The R6 community was already considering a secondary stress situation where the procedure could be considered overly conservative, but GDA needed to base its assessment on the procedure as it currently stood. In consequence, a hybrid approach was used for GDA, but a much more detailed comparison of R6 and RSE-M has now been undertaken by the licensee in the site-specific response to an assessment finding.

Overall, ONR expects conservative defect assessment calculations, but equally there should be reasonable demands placed on inspection qualification and achievement of material properties.

5. Multidiscipline Interactions

Where highest reliability is not claimed, there must be a consequences case taking account of the design provisions to maintain delivery of the safety functions. In these situations a common challenge across the GDAs relates to the integration of the structural integrity case for metal components and structures within the overall PSCR. In particular, the structural integrity case, and hence inferred structural reliabilities, must be commensurate with the UK structures, systems, components (SSC) categorisation and classification (SAP ECS.1 to ECS.5).

Thus to achieve coherency in the safety case and its assessment, the claims, arguments, and evidence needs to be integrated across the technical disciplines. From the structural integrity perspective some frequent multidiscipline interactions involve: (i) fault studies;

GDA Technical Disciplines

1. Internal Hazards
2. Civil Engineering
3. External Hazards
4. Probablistic Safety Assessment
5. Fault Studies
6. Control & Instrumentation
7. Electrical Power Supply
8. Fuel Design
9. Reactor Chemistry
10. Radiation Protection & Level 3 PSA
11. Mechanical Engineering
12. Structural integrity
13. Human Factors
14. Management of Safety & Quality Assurance
15. Radwaste & Decommissioning
16. Conventional Safety & Decommissioning
17. Security
18. Severe Accident Analysis
19. Fire Safety
20. Project
21. Generic Environmental Permitting

Fig. 2. — Generic design assessment technical disciplines.

(*ii*) internal hazards; (*iii*) reactor chemistry, PSA; (*iv*) fuel and core; (*v*) mechanical and civil engineering; and (*vi*) management for safety and quality assurance (MSQA) disciplines (Fig. 2).

Existing safety cases for the reactor designs usually include consideration of the direct consequences through linkage to fault studies—frequent and infrequent design basis failure, for example—but there can be challenges for the internal hazards assessment when gross failure had not previously been conceded. This is because other regulatory regimes may allow a default assumption that pressure equipment designed to a nuclear design code will not fail in a catastrophic manner, which leads to plant designs based on partial rather than gross failures.

As an illustration, the integration of structural integrity claims with the internal hazards discipline has proved problematic across both the UK EPR™ and AP1000® GDAs. The issues here relate to assuring consistency in the assessment of the consequences of gross failure. These consequences include the direct consequences associated with the loss of the pressure boundary (e.g. loss of the safety functions), which tends to involve interactions with fault studies, probabilistic safety analysis (PSA) and fuel and core. Whereas the indirect (or secondary) consequences involve internal hazard specialists as the concerns relate to the effects of flooding, pressurisation, environmental qualification, pipe whip, and missiles.

However, in a number of GDAs, whilst the structural integrity case is founded on a postulated gross failure, the internal hazards case may be founded on partial consequences or the leak before break (LBB) concept. To illustrate, for the UK EPR™, the design of classified pipework was initially based on a leak-only failure mode, with leak area of $D_t/4$ (D = diameter, t = thickness). To meet UK expectations there needed to be consideration of gross failures.

Similarly, for the AP1000®, LBB relates to limited consequences and involves a claim that on breaching the pressure boundary the resulting crack will be stable and readily detectable prior to growth to a limiting size. In the UK, LBB is viewed as a supporting argument and not a primary means of underpinning a structural integrity case. As a result the effects of flooding and pressurisation with higher leak rates, potential missiles, and pipe whip, in terms of both adjacent SSC and restraint design, warrant further consideration to meet UK expectations.

IV. Advanced Nuclear Technologies and Structural Integrity

1. Background

There is a growing interest in ANT as a means to secure clean energy in the UK and internationally. In October 2017 the UK government (Department of Business Energy and Industrial Strategy [BEIS]) announced that it will invest up to £7 million as part of the clean growth strategy to develop further the capability of nuclear regulators who assess ANT. ONR is gaining familiarity with these technologies and their challenges through several initiatives including training, engagement with industry, and

interactions with other international regulators. ONR has developed a programme of work to grow ONR's capability and capacity in ANT, with the focus on:
- SMRs such as light or heavy water cooled reactors
- advanced modular reactors (AMR) such as Generation IV+ types (liquid metal cooled, high-temperature gas-cooled, and molten salt cooled designs)

Note that ONR is not assessing particular designs at this stage, but gaining knowledge of the new concepts and technologies.

2. SMR Project

SMR designs are an emerging technology and are in general less well developed than commercially available 'large' reactor designs. SMRs present novel design, construction, deployment, and operational concepts that may not be currently reflected in ONR's current SAPs and TAGs. The main objective of the SMR project is to ensure that ONR's assessment guidance is fit for purpose with respect to the regulation of SMRs. ONR is currently planning a review of assessment guidance. The aim is not to review all SMR designs in detail, but to focus on key concepts, and to establish whether current guidance is adequate or whether additional guidance either in the form of new SAP and (or) developments in TAGs are needed.

ONR's review of current guidance will be informed by developing knowledge and expertise through training and engagement with the nuclear industry. The industry engagement has two principal purposes: firstly for ONR to gain an understanding of SMR design concepts and their deployment models; and secondly, to allow the industry, which includes several international vendors from prescriptive nuclear safety regulatory regimes, to gain an improved understanding of ONR's non-prescriptive approach and expectations. ONR is also seeking opportunities for multilateral working and cooperation with other international regulators. However, there are limits to how far multilateral working can progress without compromise to the achievement of the individual countries' legal duties; in the UK, for example, the duty holder is legally bound to reduce risks SFAIRP under the *Health and Safety at Work Act* 1974.

The current integrity of metal component SAP and supporting TAG have been applied to a range of relatively mature nuclear technologies in GDA. However, for the structural integrity discipline, several future challenges are envisaged in assessing and regulating SMRs. A key challenge will be to develop effective processes and approaches for the regulation of modular construction, assembly, and transport. This will be a particular challenge if the structures and components are classified by the duty holder as highest reliability. For example, the compact nature of SMRs may, if not addressed at the design stage, constrain the ability to undertake inspection during manufacture and during service. In addition, the risks and potential for damage during transport must be understood and managed. Further challenges relate to the justification and assessment of:
- novel design features such as integrated PWR designs with the pressuriser and (or) SG contained within the RPV, including classification and provisions for examination, inspection, maintenance and testing (EIMT)

- the applicability of PWR operational experience to the specific design conditions envisaged for SMRs along with, if appropriate, proving performance using first of a kind testing
- the classification, and hence establishing the structural integrity provisions, for passive and active safety systems
- the development of new materials, fabrication, and construction practices, which will need prior approval in recognised design and construction codes

3. Advanced Modular Reactors

The AMR project has several objectives. AMRs are Generation IV reactor designs that are less mature in their development than their 'large' mature counterparts. AMRs make use of novel technologies and are not based on LWR design principles. A key objective is therefore to improve and enhance ONR's capability and capacity to underpin the future regulation of AMR technologies. A further objective is to provide support to the BEIS AMR feasibility and development project, which is expected to be announced over the next few months. In support of these objectives, ONR has undertaking high-level reviews of Generation IV technologies to gain familiarity with the design concepts and their challenges. This has also served as a basis to develop the regulatory criteria and associated guidance for the BEIS AMR feasibility and development project. ONR is also planning engagements with AMR vendors who plan to take part in the AMR feasibility and development project. In addition, the high level AMR technology reviews have informed the identification of skill and knowledge gaps, i.e. areas and concepts where ONR needs to invest resource to improve understanding through, for example, more detailed review work. ONR has used the output from the high-level review of the AMR technologies to develop a training plan for Inspectors.

From a structural integrity perspective several of the challenges outlined for SMRs are applicable to AMR—for example, compact designs, EIMT challenges, and novel deployment models. As with SMRs there may also be novel design features and concepts. However, initial considerations suggest that AMR technologies are likely to present a wide range of new material and fabrication challenges: high temperatures, aggressive environments, high irradiation conditions, and sometimes a combination of one or more of these factors. The available OPEX may be limited in some cases e.g. lead-cooled reactors and molten salt reactors. Moreover, the experience base for AMR technologies founded on available civil experience, e.g. sodium-cooled reactors and high-temperature gas reactors, currently falls well short of the envisaged design life of 60 years. There is therefore a strong impetus to gain further understanding of the through-life behaviour of materials (including welds), synergistic effects, cliff-edge effects, along with risk mitigation measures and strategies. For structural integrity these challenges are significant because the duty holder not only classifies certain structures and components as warranting high levels of structural integrity demonstration, but these are often life-limiting. The duty holder's provisions for ageing management are therefore expected to be crucial to underpinning the structural integrity of AMR technologies. It follows that ONR will take a keen interest in, for example, the developments and approval of new materials, fabrication processes, and inspection techniques in recognised codes and standards as a prerequisite to their use in nuclear safety cases for AMR technologies.

V. Conclusions

GDA provides an opportunity to reduce regulatory risks at the design stage and is a multistep assessment process with increasing scrutiny. For example, for structural integrity a claims, arguments, and evidence format tends to be used. GDA is now a well-established assessment process which has been successfully applied to a range of new reactor designs. The current status of new reactor designs with respect to GDA is described. In addition, an outline of potential developments in the design review process is discussed. These developments reflect the learning and experience gained from past GDAs whilst also anticipating the future challenges associated with the introduction of ANT.

Structural integrity assessment is based on ONR's assessment guidance as detailed in the ONR SAP and supporting technical assessment guides; this guidance applies to all assessments, not just for 'new build'. A key difference with other regulatory regimes for RPs relates to the UK non-prescriptive (goal setting) regulatory regime. This presents both challenges and opportunities to reduce risks. For structural integrity some key challenges overcome by RPs during GDA are discussed.

There is a growing interest in ANT as a means to secure clean energy in the UK and internationally. ONR has developed a programme of work to grow ONR's capability and capacity in ANT. Some future challenges from a structural integrity perspective are outlined.

Acknowledgements

The author appreciates the support from colleagues within ONR and thanks technical support contractors who ably support our GDA assessments, including the late Professor John Knott, for his support to ONR during GDA.

References

1. ONR, "New Nuclear Reactors: Generic Design Assessment Guidance to Requesting Parties", ONR-GDA-GD-001 Revision 3, September 2016.
2. J. P. Caul and A. J. Holt, *in* "Materials and Methodology Challenges for Future Nuclear Power Plant" (R. A. Ainsworth and P. E. J. Flewitt, eds), pp. 14–25. EMAS Publishing, Warrington, UK, 2017.
3. P. Harrop, "UK Nuclear New Build, Generic Design Assessment, Progress with HSE Nuclear Directorates Assessment for Structural Integrity". TAGSI/FESI Symposium, "Structural Integrity Issues: Current Worldwide Trends for Nuclear Power", Cambridge, UK, 20th April 2010.
4. G. J. Hopkin, *in* "Materials and Methodology Challenges for Future Nuclear Power Plant" (R. A. Ainsworth and P. E. J. Flewitt, eds), pp. 123–26. EMAS Publishing, Warrington, UK, 2017.

5. ONR, UK EPR™ GDA Step 4 and Close-Out Assessment Reports.
   http://www.onr.org.uk/new-reactors/reports/step-four/technical-assessment/ukepr-si-onr-gda-ar-11-027-r-rev-0.pdf
   http://www.onr.org.uk/new-reactors/reports/step-four/close-out/gi-ukepr-si-01-close-out.pdf
   http://www.onr.org.uk/new-reactors/reports/step-four/close-out/gda-close-out-gi-ukepr-si-02.pdf
6. ONR, AP1000® Reactor GDA Step 4 and Close-Out Assessment Reports.
   http://www.onr.org.uk/new-reactors/reports/step-four/technical-assessment/ap1000-si-onr-gda-ar-11-011-r-rev-0.pdf
   http://www.onr.org.uk/new-reactors/ap1000/reports/assessment-reports/onr-nr-ar-16-009.pdf
   http://www.onr.org.uk/new-reactors/ap1000/reports/assessment-reports/onr-nr-ar-16-010.pdf
   http://www.onr.org.uk/new-reactors/ap1000/reports/assessment-reports/onr-nr-ar-16-005.pdf
   http://www.onr.org.uk/new-reactors/ap1000/reports/assessment-reports/onr-nr-ar-16-011.pdf
   http://www.onr.org.uk/new-reactors/ap1000/reports/assessment-reports/onr-nr-ar-16-012.pdf
   http://www.onr.org.uk/new-reactors/ap1000/reports/assessment-reports/onr-nr-ar-16-013.pdf
7. ONR, UK ABWR GDA Step 4 Assessment Report.
   http://www.onr.org.uk/new-reactors/uk-abwr/reports/step4/onr-nr-ar-17-037.pdf
8. ONR, "Safety Assessment Principles for Nuclear Facilities", Revision 0. Office of Nuclear Regulation, Bootle, Merseyside, UK, 2014.
9. ONR, "Technical Assessment Guide: Integrity of Metal Components and Structures", NS T/AST/016 Revision 5. Office of Nuclear Regulation, Bootle, Merseyside, UK, 2017.
10. ONR, "Technical Assessment Guide: Guidance on the Demonstration of ALARP", NS T/AST/005 Revision 9. Office of Nuclear Regulation, Bootle, Merseyside, UK, 2018.
11. ONR, "Technical Assessment Guide: The Purpose, Scope and Content of Nuclear Safety Cases", NS T/AST/051 Revision 4. Office of Nuclear Regulation, Bootle, Merseyside, UK, 2016.
12. ONR, "Technical Assessment Guide: Categorisation of Safety Functions and Classification of SSC", NS T/AST/094 Revision 0. Office of Nuclear Regulation, Bootle, Merseyside, UK, 2015.